# SEXUAL ECONOMYTHS
## Conceiving a feminist economics

Chris Beasley

St. Martin's Press • New York

*To Peter and Bev and Vic*

First published in the United States of America in 1994
Set in 10/11 Sabon by DOCUPRO, Sydney Australia

Printed in Singapore

ISBN 0–312–12234–9 (cloth)
ISBN 0–312–12235–7 (paper)

Library of Congress Cataloging-in-Publication Data

Beasley, Chris.
Sexual economyths: conceiving a feminist economics/Chris Beasley.
   p. cm.
   Includes bibliographical references.
   ISBN 0-312-12234-9 (hc) ISBN 0-312-12235-7 (pbk.)
   1. Sexual division of labor. 2. Sex role–Economic aspects.
   3. Feminist theory–Economic aspects. 4. Economics–Philosophy.
   5. Marxian economics. I. Title.

HD6060.6.B4 1994

306.3' 615—dc20                                                    94—9072
                                                                       CIP

# SEXUAL ECONOMYTHS

# Contents

# Introduction: a theory of feminist economics, or the flounder recants

> I can't help you, my son . . . You have misused all the
> power I gave you . . . No, my erring son, you can't expect
> any comfort from me. Your account is overdrawn. Slowly,
> a little late perhaps, I have discovered my daughters.
> (Günter Grass, *The Flounder*[1])

The title of this introductory chapter is intended to refer to two
conceptual beginnings for the book with regard to its central focus
and its approach. The first part of the title alludes to what I see as
my overall project: to provide an extended discussion of a debate
around labour and economics that reached an especially significant
point in the late 1970s to early 1980s as feminists in Western
countries directed their attention to schematic accounts of connections between class and 'gender' relations. Barrett and Delphy, in
particular, considered in their different ways the question of whether
a 'materialist' feminism was possible[2] and so raised a series of related
problems, including whether a feminist economics could be elaborated and whether a feminist epistemology of power could be situated
in relation to such an economics. However, debate amongst feminists
in the West regarding theories of labour and economics appeared to
reach something of a standstill in the 1980s. Perhaps this occurred
in part because the arrival of other theoretical frameworks like
postmodernism and post-Marxism threw many of the assumptions
employed by writers such as Barrett and Delphy into disarray.
Certainly, as the authority of Marxism waned, the means to a
feminist reappraisal of economics and power became less clear.

At the same time, the cultural milieu of Western feminists became
increasingly marked by its privileging of economic analysis. That
milieu is not limited by nation–state boundaries but is evident
throughout the Western world. The rise in the importance of economic rationalism, and its attendant dramatic effects on public policy
in countries such as the USA, the United Kingdom, New Zealand

and Australia[3], has resulted in the growing legitimacy and hegemony of the language of economics as a means for understanding all manner of social arenas. Indeed, Alain Touraine comments on this point, '[s]ocial life is now conceived only as a market and it is not surprising to see in the social sciences, which were constituted in opposition to the a prioris of economic liberalism, the return of the model of "economic man" '.[4]

The linkage of the increasing appeal of economic analysis with the reconstitution of a particular form of economics modelled around an androcentric norm is of some concern for feminists. This concern gains weight when it is noted that, despite Paul Samuelson's declared view of economics as 'the *queen* of the social sciences',[5] feminist economists have consistently suggested a majestic disinterest in women's position within the field of economics.[6] Added to this, it would appear that Western countries, in which the prevailing dominance of economic discourse is evident, are also undergoing serious economic difficulties.[7]

In light of the expanded explanatory scope permitted the language of economics, uncertainties about the capacity of conventional economics to take account of women's experience, and economic recession in Western countries, the seeming standstill in feminist debates regarding theories of labour and economics does appear to be of grave importance. It is significant not only because of the potential for feminism's marginality in conceptions of 'modern' Western social life to increase, but additionally because the problem of women's unequal economic position continues to be relatively unacknowledged and undertheorised. The problems located by writers like Barrett and Delphy in conceptualising a feminist response within the arena of economics appear all the more compelling. It seems to me that the time is indeed ripe for a reconsideration of women's labour and 'the economy'.

### Focus and limits

I am concerned in this context with how we think of economics (and power). This book deals in particular with a redefinition of notions of 'the economic' by examining the specificity of economic processes in modern Western societies from a feminist perspective as well as the possibility of a reconsideration of the concepts, categories and methods of economics. The approach is primarily analytical rather than empirical, descriptive or policy-oriented and involves developing, through discussion of several bodies of knowledge, an epistemological framework for feminist economic paradigms and notions of power. Such a focus is related to the limits of existing materials in the field of analyses of women and economics.

There are substantial bodies of feminist and other analyses in this field which largely deal with women's waged work and many approaches, especially in the area of women and development in the Third World, which consider links between women's market and subsistence work.[8] However, a far smaller number of studies focus on the 'private' labour of women in the household, or discuss at any length the connections between this labour and market work in the 'public' sphere.[9] The overwhelming majority of feminist and other literature which examines 'the economy' and the position of women concentrates upon women's public market labour and related public economic policies.[10] Thus it is severely restricted in terms of considering the specificity of women's work and the ways in which it may depart from the logic of the market. Those writers who focus on women's domestic work are inclined to note the particularities of women's labour but, insofar as they theorise about household work[11], such writers also generally accept frameworks derived from analysis of market labour relations.

Commonly employed economic frameworks ranging from 'mainstream' perspectives, such as conservative liberal (neoclassical) and social liberal/social democratic approaches, to Marxian analyses[12] are all inclined to make use of paradigms which privilege the public marketplace.[13] However, commentaries upon these frameworks, whether they arise out of discussion of women's waged work or women's domestic labour, typically just note women's exclusion from or marginality within economic categories, concepts and methods. There is often some mention of the view that the inclusion of women is likely to alter the paradigms employed, but as a rule little further comment is offered. Most frequently what is attempted in response to women's exclusion or marginality from economics is an adjustment of those paradigms.[14] An alternative perspective, a fundamental reconceptualisation which departs from 'mainstream' *and* Marxian epistemologies, is almost never expounded.

It would seem, then, that writings which examine 'the economy' and the position of women do pay critical attention to widely held 'sexual economyths' (that is, implicit or explicit economic assumptions or principles which systematically marginalise and/or exclude women) with regard to conceptions like 'women do not work', 'women are unproductive' or 'women are absent from or unimportant in "the economy" '. Women's exclusion or marginality in terms of the *content* of economics is regularly pointed out. Nevertheless, even this critique of the content of economics is limited insofar as most of the literature on women and economics deals with waged work. Since sexual economyths which assume that women are not significant in 'the economy' often do so precisely because of women's involvement in 'private' labour[15], literature considering women's waged work can only provide a partial response to the problem of content. For this reason this book intentionally focuses on the particularities of women's activities and especially on the 'private' unpaid labour they undertake.

However, such a focus is not sufficient insofar as, while there has been some critique of the content of commonly employed economic frameworks, the same attention has not been given to the *epistemological principles* underlying these frameworks. Although feminist and other literature dealing with women and economics usually notes that women's inclusion is likely to alter the paradigms used in economic analysis, such writings rarely attempt a redefinition of 'the economic' which moves beyond frameworks that implicitly or explicitly privilege the market. A challenge to market-based or derived economic paradigms is viewed as a significant aspect of any possible feminist reconceptualisation of sexual economyths, whether the paradigms employed form the basis of 'mainstream' neoclassical and social liberal/social democratic approaches which support the market, or Marxian accounts which do not.

Very few feminist or other commentators on women and economics consider such an epistemological reconceptualisation. Nancy Hartsock's *Money, Sex and Power* and Christine Delphy's *Close to Home* are two rare examples.[16] While there has recently been a spate of writings offering analyses of epistemological difficulties associated with the inclusion of women in the fields of philosophy, sociology and political theory amongst others, and hence with the question of epistemological androcentrism in these fields of analysis,[17] this mode of discussion remains highly unusual in the arena of economics.[18] Indeed Stilwell has argued generally that 'feminist analyses have made less impact in economics than elsewhere in the social sciences'.[19] Consequently a second characteristic of the book, which has already been noted, is found in its concern with the analysis of epistemological frameworks in economics. In this setting I intend initially to consider frameworks that have held and continue to hold sway in feminist accounts of modern Western women's labour. Marxism and other approaches, such as household work studies, commonly employed in relation to economic analysis of women's position provide inadequate accounts for conceptualising a feminist political economy since they adopt epistemologies which produce insufficient attention to the specificities of women's labour in modern Western societies.

Secondly, I analyse various feminist approaches to women's work and 'the economic' in order to reconfigure further an appropriate epistemology of economic processes. By working through the frameworks that have been used by feminist writers in the field, I suggest an alternative perspective and range of parameters for a theory of economics. Eva Cox has suggested with regard to the growing authority and scope of an economic viewpoint dominated by economic rationalism that critical responses to this viewpoint have remained restricted to 'an Oliver Twist cry for more social justice . . . [which has] translated into individualised or group pleas for compassion'. She insists that such pleas have been no challenge to the 'disease' of 'economic fundamentalism' that is 'rampant in the Anglophone countries', and bemoans the lack of an alternative vision of economics and social life.[20] The critique of sexual economyths undertaken in this book, through discussion of Marxism, household work studies and a range of feminist accounts, is intended to give rise to an alternative perspective that in part at least may answer Cox's call for an economic framework which refuses market reductionism and offers another viewpoint.[21]

Having outlined in brief some of the book's distinguishing foci, it will no doubt also be useful if I emphasise certain crucial limits. First, despite the significance of neoclassical liberal economics in Western societies, little direct consideration of this approach is included here. While the critique of sexual economyths and the alternative parameters developed clearly could be applied to analysis of neoclassical economics, the focus on frameworks which have held sway in feminist accounts of labour and economics effectively precludes much explicit consideration of economic orthodoxy.

Few feminist writers have employed a neoclassical approach.[22] Nor is it common for feminists even to undertake an explicit critique of neoclassical economic theory.[23] Most often feminist commentators have proceeded at a distance from this theory and have typically merely pointed out its intellectual rigidity and accompanying chilly reception of a range of issues, including 'women's issues', 'that might cast doubt on the perfect operation of the free market and the perfect justice of the results'.[24] Feminists have generally viewed neoclassical economics as, in Barbara Bergmann's words, 'hostile to any suggestion that the economic position of women was unfairly disadvantageous'.[25] Hence, feminists dealing with labour and economics have generally found little to interest them in this orthodoxy, or at most have employed only some aspects of it in combination with a range of other perspectives, thus substantially diminishing its analytic significance in their approaches.[26] Given the marked tendency to dismiss or delimit the claims of neoclassical economics, I decided that this form of economics largely fell outside the scope of the book,[27] and that exploring neoclassical parameters was in any case unlikely to lead to the development of an alternative vision. More importantly, any critique that might arise in relation to the inadequacies of Marxian and household work studies in dealing with specific forms of women's labour would undoubtedly apply to a greater extent in the case of neoclassical economics. Thus to discuss the latter seemed potentially repetitive.

For similar reasons to those raised earlier regarding the scope of the book and the coverage of neoclassical economics, postmodernist and post-Marxist approaches are not considered in any detail here. (I have discussed these approaches elsewhere.[28]) Although Western feminist thought has been greatly influenced in recent times by postmodernism in particular,[29] postmodernist and post-Marxist approaches have been much less prominent in the area of feminist accounts of economics and labour.[30] Indeed relatively few feminist writers employing, for instance, 'deconstructionist' perspectives have attended to this area.[31] Certainly such perspectives have not held sway in feminist economic analyses and hence are beyond the parameters of this discussion. On the same basis the book does not examine at any length the various forms of psychoanalytic feminism.[32] Nevertheless, it contains some references to the work of Nancy Chodorow, Juliet Mitchell and Luce Irigaray among others, and occasional discussion of the relation of the psyche to labour relations, especially labours linked to care and the constitution of the subject. Additionally, I have concentrated upon relations between the sexes and largely upon labour relations associated with heterosexuality, though I make mention of labour provided by women to women and girls. This concern with labour performed by women within a sexual social order and with heterosexuality reflects the emphasis of most feminist writings in the domain of women's work.

Finally, while I make some reference to perspectives asserted by women 'of colour' in Western countries and women of the Third World, I do not cover the field of 'women in development' (WID) studies. In particular I do not assume that the economisation of social and political questions in the Western world and the associated increasing authority of economic rationalism are inevitably a crucial problem for Third World countries or Third World women (although given the possible border crossings that occur in international divisions of labour this may be so). Rather, I concentrate upon 'modern' Western societies. WID studies may indeed have much to offer any investigation of women's work. Within the WID field a variety of works could be brought to bear on any attempt at comparing the ways in which privileging of the market, for example, may affect women in the Third World as against those in the First World.[33] However, because it may be inappropriate to generalise about women's positioning and labour on the basis of the range of experiences of Western women, there are serious difficulties associated with texts which attempt a supposedly global coverage starting from these experiences. Despite these potential problems, consideration of how to go about such comparisons is certainly a task worth attempting. Nevertheless, it is not my task in this book. Indeed, I make no proposals concerning 'developing' societies.

I assert this restriction on the scope of the book while also recognising the blurring of national boundaries linked to movements of peoples and to the internationalisation of elements of Western economies.[34] I am aware, in other words, that there is no neat dividing line between the First and Third Worlds. Moreover, as Jan Pettman has pointed out, there are potential dangers associated with too reverential a response to 'difference' between groups of women.[35] On these grounds it is evident that a comparison of understandings of economics in the First and Third Worlds relevant to women could well be undertaken, but I wish to produce a more specific analysis than this. It seems to me that a merely gestural concern with the potentially very different experiences of Third World women would not be adequate and, furthermore, that it is decidedly appropriate for Western feminists to *focus* on the particularities of cultures with which they are supposedly familiar. A central theme of this book is that texts on 'economics' and Western women's labour have assumed a self-evident place for that labour. This has even been the case in relation to feminist writings. Such assumptions require examination—that is, the seemingly familiar demands critical analysis by Western feminists as much as that which has been designated exotic.

### Terms and Language

A specifically feminist economics requires that feminism become the basis for speculation about the parameters for discussion, the significance of issues within the field of that discussion, and the compatibility or convergence of other perspectives dealing with economics. With this strategy in mind, I define feminism for the moment in theoretical terms as an epistemology which is directed towards securing recognition that women/feminine/female are as crucial an element of the human as men/masculine/male.[36] This is not to claim for feminist theory a homogeneous character but to suggest, despite the existence of internal contention, that it is possible to identify some of its broad goals, purposes and constituting categories. Jane Flax has indicated certain directions that are appropriate for this task:

> [a] fundamental goal of feminist theory is (and ought to be) to analyse gender relations: how gender relations are constituted and experienced and how we think or, equally important, do not think about them. The study of gender relations includes but is not limited

to what are often considered the distinctively feminist issues: the situation of women and the analysis of male domination. Feminist theory includes an (at least implicit) prescriptive element as well. By studying gender we hope to gain a critical distance on existing gender arrangements. This critical distance can help clear a space in which re-evaluating and altering our existing gender arrangements may become more possible.[37]

Such a definition implies no self-evident takeovers or alliances in a potential meeting with other perspectives which situate economics in social theory. Feminist theory, as described, does not insist upon supports. In common with Morris and Pringle, I contend that feminism does not in any sense *need* other theoretical positions as complement, supplement or tutor/master.[38] Within the definition provided, feminism is neither overlord nor beggar. On the other hand, feminism does not inevitably assert itself as the only or *the* central subject of theory, though within the confines of the book *a* centrality is strategically assumed.

Though Flax's working definition of feminism includes the term 'gender', debates on the use of 'gender' versus 'sex' indicate significant problems in the exclusive use of either terminology.[39] Neither term escapes its connotations regarding an assumed dualistic division between 'society' and 'biology', and hence both provide limited opportunities for displacement of disembodied social constructivism or notions of an asocial naturalised 'core' to sexuality.[40] Nevertheless I can see some advantages in employing the latter term, particularly in instances where the directly sensuous, libidinal and bodily aspects of women's experience may be commonly overlooked. This is decidedly the case in the field of economics, and therefore I have some sympathy with Hartsock's position that women's labour should be described by reference to 'sex'. She says in this setting,

> I will discuss the 'sexual division of labour' rather than the 'gender division of labour' to stress, first my belief that the division of labour between women and men cannot be reduced to purely social dimensions. . . A second reason to use the term 'sexual division of labour' is to keep hold of the bodily aspect of existence—perhaps to grasp it over-firmly in an effort to keep it from evaporating altogether.[41]

I suggest that use of the term 'sex', because it implies holding on to the bodily aspect of existence, allows an explicit challenge (in some respects at least) to the separation of the body from social meaning and its denigration in or exclusion from social and political thought.[42] Moreover, the bodily associations of the term suggest recognition of different kinds of bodies which offer different capacities to the organisation of power relations. It is perhaps particularly pertinent in a book about women and labour to employ a terminology which brings these issues to the forefront By contrast, 'gender', with its emphasis on social construction, is not as effective in drawing attention to the political material of bodies.[43] For this reason 'gender' may be viewed as a less promising term within the specific context of this book. The use of composite terminologies like 'sex/gender' could be an appropriate means to avoid the demarcation of social and biological factors, but since the composite entity retains the problems of both 'sex' and 'gender' it is not clear if its use confers any analytical benefits.

While other terms have certain drawbacks, 'sex' is not by any means an unproblematic choice. Its employment does raise the danger of uncritically constituting the female body as 'the cause and irreducible meaning'[44] of the category 'women' which Flax situates as critical to feminist analysis. However, Hartsock's reasoning suggests a strategic deployment of the term which draws upon its comparatively wider usage in English language and culture, as well as its bodily associations. It is a strategy which perceives the female body not as a fixed referent but as a 'morphology'.[45] 'Sex', in this sense, is a thoroughly political category. Thus it is possible that some men may perceive themselves as women in men's bodies, but use of the term 'sex' is intended to render politically intelligible the difference between a 'femininity' inscribed upon a female body as against that inscribed on a male body.[46] In other words, use of the term is intended to acknowledge that there is no neutral body.[47]

## Theory–making: myth's form

While I judge the problem of conceptualising a feminist economics to be of great importance to feminism, I am also aware of the vagaries of such theory-making. Hence the second part of the title of this chapter. Here I am referring to Günter Grass's novel *The Flounder*, which notes the contradictory imperatives of constructing accounts of history and social life, of the practice of theory-making itself. Grass's novel involves a reworking of the traditional fairytale about a fisherman who catches a magical flounder. The Flounder offers the man the usual 'anything you desire' in exchange for being set free. The fisherman discusses the proposition with his wife, who demands that he make increasingly outrageous wishes and finally insists on powers no less than those of God. It is a tale of woman's insatiability, of the virago's lust for power. The novel presents the Flounder's refutation of this traditional fairy story. He admits that the tale is a misogynist myth propagated to conceal *his* active intervention in support of male domination. The fish acknowledges that he deliberately set out to overthrow women's hegemony at the dawn of human history and to replace it with male authority. The myth of the voracious wife and the kindly, patient and henpecked fisherman functioned to justify men's control over women.

The Flounder apologises to womankind, saying that he made a mistake and wants to put it right by supporting women's cause. Men, he believes, have messed things up. The Flounder admits that he may have gone overboard in giving men too much power, but on the other hand argues in his own defence that he did give women some pointers towards emancipation. He is tried by a tribunal of women. His justification for his intervention in history, his place in the overthrow of women and his view that he gave women opportunities for liberation are reviewed in a far-reaching analysis of human history (mostly European). The Flounder's claim to mercy and his account of history are criticised as yet another set of male mythologies. Alternative perspectives are advanced. The reader is faced with a series of interpretations from various feminist traditions to be placed against those of the recanting (male) Flounder. One sees the process of myth-making at work.

What has this to do with a book about feminism, economics and labour? It seems to me that Grass—a male novelist—highlights the ambiguity of theoretical interpretation. The Flounder says that all previous accounts of history and society are lies, the oppressor's myth, fundamentally ideological in the pejorative sense. And yet he and the women who dispute his viewpoint must often use the material of that myth to discover other ways of seeing and analysing. The Flounder says that theory lies. It is all male fantasy. Then he too is accused of lying, of further storytelling. The women at his trial fight tooth and nail over the construction of new myths. Interpretation/theory appears in this context as a fantastical activity and yet as crucially important. It seems both everything and nothing.

Such a perspective gave me pause. This book is intended to offer a critique of sexual economyths and through doing so to propose some alternative parameters for a feminist economics. In structure, it follows in many ways Grass's depiction of a continuous investigation and reconsideration of dominant assumptions, and perhaps therefore raises similar questions regarding the possibilities and limits of myth/theory. Is it possible to conceive of this book as both another attempt at myth-making and yet worthwhile? Is it possible both to constitute a viewpoint which involves reconstruction and reinterpretation and yet not to believe in its ultimate status as Truth or Reality? Should I draw attention to the expedient polemical and practical underside of theory-making? Does it undermine the storyteller's case to admit that s/he is telling a story? The point of putting forward a particular thesis is after all to be believable and persuasive. A concern with the supposed 'realities', with the 'materiality', of economics is surely even more bound to believability. Can one express doubt about the nature of Truth, about theoretical reconception, while arguing that some concept of truth is necessary? I suggest that it is indeed possible to recognise this ambiguity without succumbing to paralysis. As Grass's male narrator says, 'stories can't help being true, but never twice in the same way'.[48] What follows is therefore *a* story which proposes a vision of women and economics that is intended to be of use to feminism. It is not less useful for recognising that the approach it offers exists in a particular moment of theory development, and has a likely limited lifespan and specific purpose.

It seemed appropriate, however, given this recognition of the mythical form of theory-making, not to proceed, in the usual way of 'non-fiction' texts, from the elaboration of a theoretical framework to the subsequent demonstration of that framework. Rather, this book attempts to take the form of a kind of detective story. I do not, for instance, provide all the definitions of terms at the beginning; instead, in the process of clarifying a series of 'clues', I work towards a 'solution'. The development of terminologies and possible frameworks for interpretation occurs along the way through critique of different theoretical perspectives. In form, the book thus resembles a funnel, which opens with a broad picture at some distance from feminist approaches and steadily narrows toward an increasingly restricted focal point. The book starts with relatively abstract questions and ends in a more practical and smaller-scale engagement with specific features of the organisation of women's labour in specific contexts. While the opening chapter on Marxism, in particular, is at times somewhat technical and hence may make for a rather difficult beginning, readers can expect to find that as the funnel narrows it also becomes more accessible.

The book opens with an investigation of the epistemological bases of Marxism, which have for some time had a special pre-eminence in feminist accounts of economics and labour, and of the assumptions of household work studies. These analyses suggest an initial account of the particularities of women's labour. The second section revolves around an exploration of various feminist views of women's labour and the question of mapping the organisation and expropriation of that labour. The funnel form of the book means that I do not return to a detailed overview which summarises the whole gamut of the text since a series of subconclusions are drawn in the process of increasing clarification and definition of the field under study. The intention is gradually to trim the various theoretical approaches in order to arrive at the story's endpoint. Consequently the closing remarks are more concerned with outlining the potential of an alternative perspective, and some further directions in that regard, than with lengthy summary.

Finally this book, in common with most other feminist accounts of theory, is not situated in one intellectual discipline but rather spreads over political economy, philosophy, politics, sociology and, of course, women's studies. It is in this, as in many other ways, a labour/'product' of its time. The scope of the story represents the delimitation of previous interpretations. Explorations of the daughters of economics must flounder at the tidemark of these interpretations.

# Part I

# Marxism, household work studies and sexual economyths

# 1 The Marx Question

> The ideas of economists and political philosophers, both when they are right and when they are wrong, are more powerful than is commonly understood. (J. Maynard Keynes, *General Theory of Employment*, Book VI.[1])

> To cite Conran's Law of Housework—it expands to fill the time available plus half an hour: so obviously it is never finished.' (S. Conran, *Superwoman*[2])

Marx's work has been of critical significance in feminist writings on 'materialism', economics and work. For this reason an analysis of his approach is an obvious and useful place to begin discussion of the possibilities of conceiving a feminist economics. However, I do not intend to provide a thoroughgoing analysis of Marx's writings, nor of the various traditions of Marxism, as such detailed exegesis has been undertaken by countless others, including many feminists. Rather this chapter and the following one offer some brief considerations on the problematic nature of Marx's proposals concerning labour/production/economics. The chapters are intended to provide an illustrative guide to aspects of the work of the 'father' of political economy as a way of developing some potential parameters for understanding modern Western women's labour.

## Economic determinacy and the 'base/superstructure model'

Initially the most outstanding feature of Marxism is the significance it accords 'the economic'. Indeed the privileged status of economics, associated with its placement in a paradigm combining generality with determinacy, may be viewed as critical to how one defines the limits of Marxism. In this setting Eagleton has suggested that while it is quite possible within Marxism to reject any number of versions of the 'base/superstructure model' of social relations, that model— which retains a conception of the primacy of economic causality—remains a linchpin of Marxist thought: '[t]he key point

3

for Marxism . . . crystallized in the base/superstructure doctrine, is that . . . [social] determinations are not symmetrical: that in the production of human society some activities are more fundamentally determining than others.'[3]

Many commentators have suggested that the employment of the base/superstructure model in Marx's work, and in Marxism generally, produces a paradigm compromised by economic reductionism. For some Marxist writers this difficulty, though regarded as evident in Marx's earlier works, is resolved in later writings.[4] Other critiques, however suggest that the concept of economic determinacy in Marx's works contains an immanent potential to collapse accounts of social life into simplistic analysis or even a reductive essentialism. Johnson, for instance, notes that Marx continued to assert that revolutionary consciousness develops in a relatively uncomplicated and straightforward way out of the economic contradictions of capitalism, despite the explication in his later works of a theory of mystification.[5] Post-Marxists such as Laclau and Mouffe go further than this and argue that all variants of Marxism, no matter how sophisticated, slip inevitably into economism because of the use of the base/superstructure model. They assert that since the critical defining feature of Marxism lies in its assertion of economic causality in the midst of a plurality of social processes, economic reductionism in Marxism is unavoidable.[6] There is not sufficient space here to discuss in detail the debates concerning the impact of the concept of economic determinacy within Marxism. Instead I will consider one of several theoretical problems that might be cited in support of accusations regarding economism, for, while a feminist economics might employ the notion of an economic 'base' which is accorded determining primacy, it must at least consider claims linking this notion with an inevitable economic reductionism.

Commentators who have drawn attention to the issue of economism in relation to Marxism have generally asserted that Marx's view of the differential status of economics implicitly accords a more or less ontological efficacy to the 'base'. In this regard Marx is accused of a tendency to confuse an ontological doctrine concerning social being and consciousness and a conception of sociohistorical analysis entailed in the base/superstructure model. As Eagleton notes in commenting upon the work of Raymond Williams, the ontological doctrine promulgated by Marx in the 'Preface' to *A Contribution to the Critique of Political Economy* is stated in the same breath as the base/superstructure claim. Eagleton points out that the two perspectives are of a critically different character, despite being pursued simultaneously in this work, and that this congruence creates serious difficulties for Marx's approach.[7] The association of the two encourages commentators' inclination to mistake the one for the other.[8]

However, as Eagleton acknowledges, such confusion is not to be placed entirely at the feet of misled commentators. He suggests in his consideration of the 'Preface' that the ease with which this confusion occurs is partly due to the close connection between the ontological and base/superstructure arguments which can indeed promote economic reductionism. The ontological doctrine simply presupposes a broadly materialist perspective in which social being determines consciousness, 'consequent upon the material structure of the human body, the material nature of its environment, the necessity for a mediatory labour between the two, and the fact that consciousness is therefore always in the first place . . . "practical consciousness" '.[9] Eagleton states that the breadth and ambitious scope of this claim is ironically more readily accepted than the base/superstructure proposal and thereby by association illegitimately justifies the latter, even though there is no 'logical entailment' between the two.

This point brings to light in his view not only a dangerous slippage in Marx's work which catalyses an assumption of unquestionable privilege and causality associated with 'the economic', but paradoxically also shows the weakness of the case for that assumption. The ontological doctrine cannot provide the foundation for the base/superstructure model, and yet Marx seems to imply precisely such a linkage. The confusion between the two formulations is of no small moment, for if it is accepted that there is no logical connection, then the hesitation of the post-Marxists over the base/superstructure model becomes rather more understandable. Post-Marxists appear not so much at odds with Marx over his ontological assertion of a broadly materialist framework for social relations, as unpersuaded by the conception of privileged determination within those relations. Eagleton's analysis suggests that they have quite correctly judged the base/superstructure model to be at least weakly established by Marx.

> By asserting the two [the ontological and base/superstructure formulations] in the same breath in his Preface, Marx runs the risk of making the latter sound like an ontological claim too, or at least making it sound as though the two go naturally together. To accept the base/superstructure argument then comes to seem like accepting, almost on faith, that this is just the way history works, just as we accept the being/consciousness case because this is just the way human animals are. But this surely obscures the historical specificity of the base/superstructure model. The question we need to pose, and which is left unanswered by Marx in his Preface, is *why* certain political, legal and other institutions, and certain definite forms of social consciousness, 'rise . . . on the real foundation' of a mode of production. To put the question another way: what is it about the

mode of production which historically (rather than ontologically) *necessitates* such products? For it is not, of course, 'economic activity' which Marx claims gives rise to such superstructures; it is economic activity conducted within relations of exploitation.[10]

It is precisely this view—that the base/superstructure model, if it is to be accepted, must rest upon the claims of history—which is shared by post-Marxists. The post-Marxist critique suggests that the model fails *because* the 'superstructures' cannot be seen as determined by economics *throughout history*, no matter how mediated the determination. Eagleton, by contrast, goes on to propose that the Marxist paradigm can provide both a logical and historical defence of the model. Asserting that economic relations happen to have been exploitative for all of human history thus far, gives history the unitary identity which the base/superstructure position requires. The model is hence made historically contingent yet ubiquitous until now. Ontology, in other words, becomes mere ubiquity. While this is certainly one way of dealing with the post-Marxist view of the illegitimate slippage between ontological and historical frameworks in Marxism[11], there remains some basis for the post-Marxist assertion that the privilege accorded economics is problematic and that the unconvincing character of that privilege is in part found in the contribution of Marx himself.

The issue of confusion between general formulations of historical materialism (ontological principles) and analysis of specific historical periods is not only relevant to post-Marxism, however. It is also of some significance in developing a feminist perspective regarding economics. Nicholson states that Marx provides an inherently inconsistent account of critical terminologies including the 'economic', 'production' and 'labour'. She postulates that this inconsistency results in an illegitimate conflation of general cross-cultural claims with particular historical settings and that a specific case of economic determinacy is thus mistakenly expanded to embrace the whole of human history. Nicholson argues that at some points 'production', for example, refers to all activities necessary for human survival, yet this broad meaning is constantly interchanged with a limited one referring only to those activities geared to the creation of material objects or, even more narrowly, only those which result in commodified objects. Such ambiguity has the serious consequence of enabling Marx to project both the primacy and the autonomy he accords the economic in capitalist societies across cultures.

This point is illustrated by examining Marx's claim that 'the changes in the economic foundation lead sooner or later to the transformation of the whole immense superstructure'. This claim is intended as a universal claim of social theory, i.e. it is meant to state that in all societies there is a certain relation between the 'economy' and the 'superstructure'. If we interpret 'economy' here to refer to 'all activities necessary to meet the conditions of human survival', the claim is nonproblematic but trivial. More frequently, 'economy' is interpreted by Marx and Marxists to refer to 'those activities concerned with the production of food and objects'. Here, while the claim ceases to be trivial, it now contains certain problems as a cross–cultural claim. While all societies have some means of organizing sexuality and child-care, it is only in capitalist society that the former set of activities becomes differentiated from the latter under the concept of the 'economic' and takes on a certain priority. Thus by employing the more specific meaning of 'economic' in his cross-cultural claims, Marx projects the separation and primacy of the 'economic' found in capitalist society onto all human societies . . .

Marx, by giving primacy to the 'economic' . . . appears to be making the . . . claim that the ways in which we produce food and objects in turn structures the manner in which other necessary human activities are performed. But the force of this latter claim, I would argue, rests upon a feature true only for capitalist society: that here the mode in which food and object production is organized to a significant extent does structure other necessary human activities.[12]

One does not have to accept Nicholson's assumptions regarding economic determinacy in class relations, the priority of capitalist economic activities over those in other spheres, her view of capitalist society (taken as meaning modern society as a whole), or her claim that determinacy is only dubious when applied cross-culturally. These positions can be considered questionable, yet Nicholson's points regarding the base/superstructure model and the autonomous character of the economic may still be acknowledged. She indicates that to the extent that Marx conceives the 'economic' as synonymous with his perception of the characteristics of the capitalist market, the base/superstructure model issues from a crucially unpersuasive ambiguity, a logical and philosophical confusion. In this light Marx's claims to a general theory of history may be regarded as at least problematic and as unwarrantably proposing an economic determinacy which is in a quite fundamental sense reductive, that is, at minimum culture-specific. She implies that Marx's inclination to describe a distinction between base and superstructure and to construe economics as a discrete realm is similarly a culture-specific reduction projected into an ontological framework. Even if Marx does not propound a complete distinction or an entirely discrete status to economics in the terms Nicholson appears to outline,[13] the notion of *a* distinction and the particularity of economics in Marx's work can certainly be viewed as a specific feature of capitalism that would seem rather more difficult to perceive in other fields of economic activity within 'capitalist society', let alone cross-culturally.

While it is possible to distinguish, for example, between the arena of the capitalist economy structured according to the commodity form and other arenas such as the state which are not organised in this way,[14] such a distinction between 'economic' and 'non-economic' cannot be easily transferred to the 'household economy' within modern societies any more than it can be straightforwardly applied to 'pre-capitalist' societies. Indeed the context of the household economy shows very clearly the reductive limitations of Marx's conflation of general theory into a notion of economics = production in the capitalist market. Relatively conservative accounts of the parameters and activities of the household economy in modern Australia suggest that this conflation is seriously inadequate.

The evidence from the 1974 survey of time use in Australian households is that paid work in the market economy absorbs *less than half* of all labour time used for production of goods and services; just *over half* is unpaid time used for household production . . . Time use surveys indicate that about 75 per cent of all time put into household productive activities is contributed by women.[15]

It is not evident why Marx assumes that the mode in which certain kinds of production of food and objects occurs should determine these other 'productions', as well as all other necessary human activities. As Nicholson notes, Marx's explanation for the premise of a particular determinacy rests upon a confusion with ontological claims, despite the considerable significance of activities which even he should regard as labour. There is no obvious reason for the exclusion of these activities from his conception of determinacy barring the confusion which she points out and its attendant implications regarding Marx's masculinist stance. No wonder, then, that many feminists remain sceptical 'about how fully "economic" arguments can explain gender oppression',[16] especially when the 'economic' is interpreted in this narrow fashion.

There are at least two outcomes of Marx's weakly defended account of 'economic' determinacy. The notion of generalised economic determinacy might be regarded as unworkable per se, and the category of economics employed by Marx would certainly appear to require some rethinking for a feminist analysis. One might assume that such a viewpoint would be grist to the post-Marxist mill in that it clearly strengthens the case for the perceived flaws in the base/superstructure model and the link between that model and economic reductionism. However, post-Marxists such as Laclau, Mouffe and Touraine, along with neo-Marxists such as Offe, appear inclined to accept Marx's equation of economics/production/labour with the operations of the capitalist market economy.[17] Post-Marxists appear on the one hand to move a long way from the Marxist paradigm, yet from a feminist perspective do not perhaps move far enough in some respects. If Marx's premise of determinacy is in part compromised by a failure to consider labours predominantly undertaken by women (in modern societies), his approach can be said to share that stumbling block with writers in the post-Marxist tradition.

## Feminist responses to the Marxist account

Even if difficulties in the post-Marxist approach are acknowledged, it is not hard to see that certain parameters of its critique of Marxism may be appropriate to feminism, including the rejection of economism and overly general (axiomatic) causality. As Eagleton notes, Marxism does not merely entail a broad acceptance of general causality. It promotes a singular account of *a* privileged cause in the last instance, since among a multitude of social determinants one has status above others, thus narrowing the category of causality considerably. At the same time Marxism enlarges the field of generality not just to certain aspects of sociality in certain periods but 'to an object as immense as history itself'.[18] Eagleton himself suggests that the scope of general causality invoked appears dangerously doctrinaire and simplistic:

> [i]s it not grossly implausible to believe that, in the end, one set of determinants alone has been primarily responsible for the genesis and evolution of forms of social life? . . . It is its historical reach which appears most dubious . . . [T]o imagine that [history] is always ultimately determined in the same way seems to ascribe to it a spurious sort of self-identity.[19]

Pondering this self-identity which post-Marxists so ardently repudiate, Eagleton concludes that it is not the notion of a primary causal mechanism per se which seems so unlikely but rather its 'overweening conceptual span'.[20] While feminist accounts of economics might affirm the possibility of employing notions of singular causality and also a degree of generality, they too may baulk at the complete knitting together of these concepts into a singular general causality to describe and explain all human activities throughout history and in all cultures. However Eagleton asserts that such an overweening span in Marxism derives from the perspective that the self-identity of history is associated with its unity as 'pre-history'.

> History has not even started yet. All we have had so far is the realm of necessity—the ringing of the changes on the drearily persistent motif of exploitation . . . What all historical epochs have in common is that we can say with absolute certitude what the vast majority of men and women who populate them have spent their time doing.

They have spent their time engaged in fruitless, miserable toil for the benefit of others.[21]

Though this claim of consistent subordination is indeed awesome, it has not blinded feminists to the problem that 'exploitation', 'the realm of necessity', and 'fruitless, miserable toil' have often not included women, that Marxism's paradigm of *exploited* labour has in fact largely referred to labour activated by male subjects and, more specifically, male subjects within Western capitalist societies.[22] Nicholson, among others, demonstrates that the 'reality' of human suffering crucial to Marxism has been for the most part a picture of men's experience of misery, caused by men and interpreted by men.[23] Of course one could attempt more thoroughly to include women in the paradigm of exploitation and/or extend exploitation to a concept of oppression, thereby apparently including them still further. But without alteration to the cornerstone of 'exploited' labour as it is understood in Marxism, the process of inclusion remains delimited and something of an afterthought. Eagleton does not sufficiently address this issue and appears, despite his interest in the inclusion of women, to see the Marxist definitions of and associations between 'exploitative' production and determinacy/causality as finally unproblematic: 'what has historically preponderated in "social being" has in fact been exploititative [*sic*] economic production'.[24] Feminists and others may wonder rather more at the assumption not only of labour as the primary determinant of sociality throughout human history, but also the assumption of privilege concerning what amounts to a particular set of labours in a particular set of organisational frameworks (largely masculine labour associated with the public production of food and objects within class relations).

Despite these problems, a number of feminist writers have drawn upon the Marxist account of economics and economic determinacy and have consequently, in common with the assumptions of that tradition, equated economics with the capitalist market and accumulation. For example, Juliet Mitchell, who proposes a 'dual systems' model of distinct patriarchal and capitalist relations, offers a perspective in which patriarchy is defined in largely non-economic terms while capitalism is described as an economic system. This view is evident not only in her early work *Psychoanalysis and Feminism*, but in later writings on the psychic organisation of patriarchal forms.[25] Mitchell does not outline economic processes specific to the relation of the sexes (in contrast to Christine Delphy's approach, for example).[26] Nor, however, does she assert that economic determinacy should be considered the motor of all social processes. Thus she allows dynamic effectivity to both sides of the dual systems model she describes.

Michele Barrett, by comparison, objects to Mitchell's and others' use of the term 'patriarchy' and the notion of historically distinct dual systems because she considers patriarchy an idealist (non-economic) conception. On closer examination, Barrett's concern appears to arise out of an assumption that the material determination of the conditions of social existence must necessarily be seen as residing in class relations. Patriarchy is ahistorical, and therefore idealist, because it is not synchronised with or related to Marxist 'modes of production'. Barrett's tendency to reject Delphy's account of patriarchy as an economic (production) system similarly reveals an unexamined privileging of analytical space for class relations. Yet there is no obvious reason why 'historical materialism' or economic analysis should be seen as the monopoly of Marxism or class theory. Additionally, from Barrett's own materialist standpoint, her refusal to concede an economic organisation to patriarchy must logically marginalise the claims of feminism as a social movement and, unlike Mitchell's account, counter the granting of central significance to sex relations in the construction of social life and history. Barrett's commitment to Marxism means that her failure to conceive sex relations as 'material' is then translated into a critique of patriarchy as idealist.[27]

Mitchell and Barrett share, despite their differences, a debt to Marxism which results in the acceptance of an equivalence between economics, the Marxist account of 'production' relations, and 'materialism'. In Mitchell's case these equations lead to the constitution of a non-economic account of sex relations; in Barrett's work they lead to a rejection of that account and of frameworks which do propose economic processes based on sex relations. In both instances there is a reinstatement of the significance of capitalist economics. These feminist writings demonstrate, as does the work of post-Marxists, that it is possible to retain questionable Marxist assumptions concerning economics in the midst of critiques of Marxism as the total explanation of social existence. This retention cannot but seriously impair the effectiveness of the critique and the alternative proposals put forward, including proposals regarding economics and economic determinacy. My dispute with feminist writers like Mitchell and Barrett, as well as with post-Marxists such as Laclau and Mouffe, involves a view that conceptions of a feminist economics in particular cannot rest upon any notion of an identity between economics/production/labour and the Marxist version of these. Clearly in this setting the assertion of a non-identity between economics and Marxist 'production' relations requires some further elaboration.

The problematic nature of any framework which accepts Marx's usage of economics/production/labour has been briefly demonstrated in discussion of Nicholson's critique, but many feminist works from a variety of perspectives have countered the Marxist privileging of class (and relatedly its particular version of overarching 'economic' determinacy). Such writings range from Elshtain's 'communitarian, even traditionalist' analysis [28] to radical, postmodernist and socialist feminist approaches. Socialist feminists like Jaggar, for instance, have disputed the Marxist tendency to conflate labour performed by women in the private sphere with the operations of waged work and the associated de–emphasis upon men's involvement in subordination which is almost eclipsed beneath the imperatives of capitalism.[29] Women's private labour is thereby reduced to a class phenomenon. But, says Jaggar, despite similarities, this labour is *not* identical with that of waged workers: the oppression of wives and prostitutes is not to be confused with the exploitation of wage labourers.[30] Hartmann likewise argues that the Marxist account of economics/production/labour will simply not do for sex relations. Hence she regards Zaretsky's view that capital created the public/private distinction and the *appearance* of women working for men (whereas in *reality* women labour for capital) as theoretically and politically misdirected.[31] Radical feminists and feminists influenced by postmodernism have also often been deeply antagonistic to any notion of the primacy of class relations over those of gender. Writers like Allen and Campioni and Gross, for instance, argue that there can be no pre-given demand that history and society be conceived in terms of Marxist historical materialism: Marxist accounts of economic determinacy cannot be seen in their view as *the* explanation of social relations throughout history.[32] Even Barrett states forcefully that the subordination of women is not pre-given or logically determined by capitalism.[33] Yet while many feminist writers broadly insist that Marx's account of class and economics is inadequate when describing the labour of Western women—let alone other aspects of their positioning—and that consequently economics is not to be equated with the Marxist depiction of this category, it is still evident that a close examination of the limits of the equation has rarely been attempted.

At a simple level one could note initially that Marx's presentation of 'labour' tends to lack particular dimensions. Labour for Marx is oddly disembodied, despite an overall emphasis on muscularity.[34] It is not sexually specific or sexually differentiated for the most part, nor is it libidinal, even if occasionally procreative. Though for Marx labour is muscular and cognitive, he describes consciousness in rather limited terms and his analysis recognises unconscious processes only to a restricted degree.[35] While for Marx the subject/body is politically passionate, it is peculiarly bereft of an emotional life. The psychic, personal and emotional features of labour are not fully incorporated into the framework and are repeatedly compromised by the concentration on the supposedly depersonalised effects of 'production' under capitalism. Such an outlook may involve problems for analysis of class relations.[36] What is more obvious and more relevant in the context of this book is that Marx's view of labour is probably singularly unhelpful when considering labours which are undoubtedly constructed in a domain enmeshed with bodily, psychic, libidinal, emotional and deeply personalised meanings.

Delphy comments that modern Western women's private labour acquires its particularity because it is furnished by personalised interrelationships of social and biological association and, one could add, care.[37] Initially, then, without any detailed analysis, one can propose that the specificity of Western women's labour within sex relations (which precisely indicates the problematic nature of the assumption that economics can be tied to the Marxist account of 'production') arises partially from its differential *site* (though this is not always in the private sphere), the *categories of human beings* deemed responsible for and who perform the labour[38], its lack of *waged* reward (it may however be rewarded financially), and clearly in relation to the *character* of the labour undertaken. Unlike 'production', Western women's labour is invariably 'interwoven with interpersonal relationships' it is work done by women for others and interpreted as a 'labour of love'.[39] In this complex interweaving of the creation of services/goods with the expression of love/affection/care one sees the formulation of an *emotional economy* which *cannot* be reduced either to Marx's narrow definitions of labour as (a) 'production' of food and objects, or (b) 'production' of commodities under capitalism, or even to (c) his broader definition of all activities necessary to human survival, since 'activities' still tend to be described with little reference to, for example, invisible, emotional and psychic aspects of labour.

This must make one uncertain about Marx's categories of economics and economic determinacy within feminism. It also promotes caution in assessing the propriety of employing Marxist content *or* methodology (category definitions and relational contexts) with regard to economics in the manner of many feminist writers, such as Ferguson, Smith, O'Brien, Hartsock, MacKinnon, Young, Jaggar, Hartmann, and Delphy.[40] I will examine these points in greater depth in the rest of this chapter and in later chapters on Marxism and 'materialist' feminisms.


## Marxism and the limits of the standpoint of exchange

If one begins from the personalised character of the labour performed by Western women within the sexual order, it is evident that this labour is undertaken on the basis not of 'exchange' in the Marxist sense but rather on the grounds of 'love' or 'altruism'.[41] Though as Delphy has indicated there may be elements of contractual exchange in such labour[42], the contract–exchange model may well be an inappropriate language for describing its character even as a metaphor. The Marxian version of exchange focuses upon (hidden) relations of dominance/subordination beneath a surface egalitarianism and upon the features of coercion and consent. However, while the notion of masked oppression has been employed by many feminists including Delphy and MacKinnon, it may be much more difficult to locate what constitutes oppression/exploitation and differentiate coercion from consent in an emotional economy. The methodology used in Marxism may simply not be applicable in an economy founded on 'love' since the apparently 'weak links in the social and cultural chains that bind women' may be intertwined with those that women themselves are loath to break.[43] The 'Women in the Home' project of the Victorian Women's Consultative Council discovered in this context that although child-oriented activities were a source of considerable unsatisfying labour, these also constituted a major site of enjoyment and satisfaction.[44] Oppression/coercion and consent/'choice' seem to be so intermingled as to defeat any simple description of relations of domination/subordination.

This is not to suggest that Marxism does not allow for complexity in its analysis of labour but it may indicate that somewhat different conceptual tools are required for an economics (and politics) which is so deeply infused by gratification and self-identity. Indeed the Marxist approach tends to rely, no matter how complex the theoretical framework, upon a relatively sharp distinction between and separation of, exploiter and exploited, upon a perception of 'outside' (external coercion) and 'inside' (internal resistance). It displays much less concern for relations of power within labour in which outside–inside distinctions are blurred and domination is not only external but is found in one's own self. The significance of sexual identity, of the subject, in the private labour performed within sex relations obfuscates the subject/object binary largely assumed in Marxist interrogations of class relations. In personalised as against relatively depersonalised labour relations, subjectivity and subjective 'investments' assume a centrality that makes it difficult to hold to Marx's assumption that exploitation/oppression is to an extent at a distance and that this distance bespeaks its inauthenticity. Whether or not the assumption can be said to hold for class relations, the spatial and visible (quantifiable) exchange model is of uncertain use within sex relations. In particular, the 'emotional labour' aspect of women's work within the private sphere may be passed over in the concepts and categories of such a model. As Hartsock has pointed out in a related critique of the continuing domination of exchange or 'market' models of power in the social sciences, there are dangers in Marx's adoption of what amounts to the 'epistemology of capitalism—the view that knowing how society works means knowing how things and people are exchanged—while attempting to discredit it'.[45] The epistemology involves slippages which in Nicholson's view are of concern for class analysis—she asserts Marx's projection of the features of capitalism onto all history, that is, his inability to escape the perspective of capitalist society—but undoubtedly such slippages raise further questions when applied to sex relations. A focus on exchange, on use and quantity rather than on meaning/quality, suggests therefore not merely doubts with regard to the Marxist paradigm. The employment of Marxist methods in, for example, the exchange perspective of feminists like Rubin and Delphy must also be closely scrutinised.[46]

Consideration of the outside/inside distinctions that underlie the exchange model imply that Marxian conceptions like alienation should be approached with caution. MacKinnon notes that there are similarities between the experience of waged workers and women's private labour: '[a]s the organised expropriation of the work of some for the benefit of others defines a class—workers—the organised expropriation of the sexuality of some for the use of others defines the sex, woman'.[47] Nevertheless she implies that the externality characteristic of the alienation of waged workers from the products of their labour, both in the externalised domination that creates this alienated form of labour and the process of externalising objectifications of labour, is not as straightforwardly evident in the case of sex relations. In the latter, MacKinnon asserts, women themselves are the objects, the product: '[w]omen have not authored objectifications from which they can be separated. The term "alienation" does not apply to women as women'.[48] Whether or not one agrees with the degree to which MacKinnon's own position remains conjugated by a Marxist framework, her conflation of sex relations with sexuality or her tendency to assume that waged workers are men,[49] she does indicate ways in which the understanding of power and economics in Marxism may be problematic for feminist appraisals of Western women's labour, particularly private labour. Once again there are critical spin-offs from such a perception. For instance, Jaggar's proposals concerning a 'unified systems' theory, in which she identifies alienation as the central concept that can accommodate both Marxist and feminist thought[50], begin to look shaky in this context.[50] One might additionally acknowledge that women's experience—their private labour most especially—is characterised not only by a comparative lack of separation (alienation) of the 'producer' from the 'object' when women are themselves the object, but also shows this ambiguity insofar as the oppressor (expropriator) may be 'the object' as well. In an emotional economy women labour upon their 'masters', thus unsettling the more clear-cut distinctions and distance between the subject/object that Marx describes in the exchange model of class relations of dominance/subordination. When the expropriator is himself the 'objectification' of labour, when his very self denotes the enactment of women's relentless labour of activation/construction of subjects amongst others, the inside/outside spatiality and (quantifiable) measurement of alienation and expropriation become uncertain indeed. This is not to propose that 'alienation' and 'expropriation' are not concepts that can be employed by feminism. Rather they cannot be immediately identified with their meanings in Marxism.

It is not evident that sex relations are equivalent to an exchange model, to an exchange (expropriation) of things/objects, for neither women nor men and children appear to be entirely 'thingified'[51] in the process of women's private labour nor is that labour's outcome entirely removed from its authors. What is perceived within the term economics as delineated by Marxism tends to revolve around separations and discrete elements, even if not the discrete individuals of Liberalism. Hintikka and Hintikka argue that Western philosophical thought has overemphasised ontological approaches which identify and individuate parts of the world and that this represents a distinctly masculine mode which may be less than appropriate to women's experience and activities: 'women are generally more sensitive to, and likely to assign more importance to, relational characteristics (e.g. interdependencies) than males, and less likely to think in terms of independent discrete units'.[52] John Fowles makes a similar point in *The Magus*: '[t]hat is the great distinction between the sexes. Men see objects, women see the relationships between objects. Whether the objects need each other, love each other, match each other. It is an extra dimension of feeling we men are without.'[53] I am suggesting here that while Marx's work is strongly 'relational' in *certain* respects, it is not relational in the sense described above. Hence such comments do draw attention to aspects of Marx's work which reflect an unintentionally masculinist standpoint even in the most abstract categories and concepts, including the approach to economics.

## Marxism, expropriation and the problems of temporal measurement

The focus upon externality, relatively sharp outside–inside divisions, discrete elements, and empirical, visible, technical, and especially temporal measurement which demonstrates this masculinism, and thus the limits of Marx's economics for feminism, may be illustrated by looking more closely at expropriation. Marx's description of and formula for assessing surplus value acknowledges the unique character of 'the labour process' under capitalism but, as Nicholson has recognised, the depiction of labour/production/economics regularly slips between this acknowledgment and more general claims. Yet the Marxist account of appropriation is not easily transferred to other labour relations of dominance/subordination. If one looks at the mechanisms that are said to affect the *rate* of appropriation—the length of the working day, the 'intensity' of labour, and the 'productivity' of labour[54]—it is clear that the analysis of appropriation rests heavily upon measurable temporality, and to an extent upon technical/organisational 'efficiency', again assessed in terms of time. Overall increases in labour time, increases in expenditure of labour performed in a given time, and alterations in the labour process (in instruments or mode of working) which render labour more efficient in shortening labour time are said to indicate an increased rate of appropriation (exploitation). Marx notes cautiously that these factors cannot be read in isolation from 'the relative weight thrown into the scale by the pressure of capital on the one side, and the resistance of the worker on the other'[55], that is, from class struggle. In this approach it would appear that the externalised 'objective' estimation of measurable economic expropriation is tempered by political struggle. Hence Griffin disputes Braverman's inclination to view Marx as postulating an inexorable logic to capital. But even if Marx does not promote such a perspective, the framework does involve a perception of economics in terms of an empirical fixity which is *rendered* contingent *by* politics. The dull clarity of economic relations, though enlivened by politics, is accessible through measurable temporality.

Expropriation is unlikely to be as clear-cut or 'objectively' measurable in sex relations, nor is it obvious that time is an adequate measure of expropriation in this setting. The length of the working day in an emotional economy is hardly a useful guide to expropriation or at least is only a broad criterion, since the labours of most mothers, for instance, never end. Moreover nurturing, even of adults, is interwoven with the whole of life. There is no 'working day' that can be separated out from non-working time. Such a distinction involves a manifest definition of work and non-work and expropriated/non-expropriated labours that is simply not straightforward when applied to women's activities. It does not yield a picture of how women experience, organise, define, or perceive their work. 'Intensity' and 'productivity' of labour are also difficult concepts to transfer to sex relations. Women may sometimes deliberately 'intensify' their labour, while simultaneously decreasing their 'productivity', by involving children in household work/activities. This work represents a considerable challenge to any simple theory of labour as well as to Marxism, since 'the cost in time and effort of getting the work done often exceeds the value of the labour finally contributed'.[56] The meaning and the social relational significance of mothers working alongside their children seems more crucial than either measurement of the amount of work done or the time taken to perform it.[57] Jacquette's opinion that the private sphere is ordered not in spatial or temporal terms but by emotional meanings seems highly appropriate in this context.[58]

'Productivity' in the above context may look rather different from Marx's concern with tools or 'efficient' modes of working, nor is the relationship between improved 'productivity' in either the Marxian sense or in the sense of emotional meaning self-evidently related to expropriation within the relation of the sexes. Technological advances in the tools of labour do not always seem to produce a shortening of labour time in women's private labour. For instance, technically sophisticated children's toys may require lengthier involvement by mothers. Similarly, 'improved' cleaning materials may be presented as 'time savers' but also promote ideals of cleanliness and hygiene that may give rise to an increase in the time given to such work. More obviously, the emotional aspects of private labour and childcare are not particularly affected by new technologies and to a large extent may be incapable of being thoroughly technicised. Consequently Meredith Edwards asserts that the demands of childcare, for example, are such that 'despite new technologies, the total *time* spent by home managers on housework over the last century has not changed' [*emphasis added*].[59] If time use does not substantially alter over long historical periods, does this mean that expropriation is unaltered? Does this say anything of importance about women's 'productivity'?

Uncertainties regarding time measurement and 'productivity' only add to the problems of making use of Marx's conception of the significance of the 'forces of production' (instruments/tools/technologies) in historical change. Certainly G. A. Cohen's defence of Marxist 'orthodoxy', of the primacy of these forces over social and cultural relations[60], cannot be upheld as a general historical theory. The specificity of economic processes within sex relations challenges the generality of Cohen's approach. This is particularly the case given that the forces of 'production' in women's private work are at least partially different from those in class analysis. The example of the special characteristics of the sexualised body is useful here. In this context Marx outlines a general definition of the attributes of the labour process which includes purposive *activity* (work), the *object* upon which work is performed, and the *instruments* employed. The labour process involves therefore, activity, via the use of instruments, which results in an altered object.[61] Though Marx mentions the sexualised body, he does not fully investigate it as an instrument of labour that differs from other 'tools'. The sexualised body, like the labourer under capitalism, can produce an altered product (value), but also combines within it notions of activity, raw material, instrument *and* 'object'/'product'. This 'tool' can never be completely separated or distinguished (unlike the labour power of the waged worker) from any aspect of the labour process in which it is employed. Action, raw material, instrument *and* 'product' remain inextricably interlocked in this instance. The conception of dividing off the instrument-like properties of the sexualised body·in order to demarcate these as forming an aspect of the technological motor of history arbitrarily distorts the specific qualities of women's private labour to the point where the conception becomes nonsensical doctrine.

Apart from doubts about notions of 'productivity' (efficiency) in the sense of 'forces of production' influencing the rate of expropriation, it is not even clear whether Marxist assumptions concerning the 'mode of working'—the organisation of the labour process and productivity/rate of expropriation—are similarly relevant to sex relations. The notion of 'efficiency' central to 'economic' theory—whether defined in mainstream terms as value-free frugality, good management and profitability or in the Marxist sense of intensified expropriation—is quite difficult to apply to the relation of the sexes. The emotional economy is not exclusively or largely driven by themes of scarcity and competition central to capitalist notions of the 'nature' of economics. Though the emotional economy is male dominated, the 'logic' of its operation paradoxically does not share the capitalist *and* masculinist orientation of these themes.[62] Household labourers or units, for instance, are not in direct competition with one another. If a housewife completes her tasks more 'efficiently' using technological devices and 'efficient' organisational principles—if in mainstream/Marxist economic terms she takes less time—she does not gain a competitive advantage or necessarily greater financial reward. Further, 'inefficient' households/household labourers do not as a matter of course consequently fail to 'sell' their 'products' or 'sell' them at a lower 'price'. Even in the case of a 'product' such as the male-waged labourer whose labour power is literally sold in the capitalist marketplace, that product does not bring in a lower wage (price) when his wife is 'inefficient'.[63] Fluctuations in 'production'—'overproduction' and 'underproduction' —however this is understood, do not inevitably affect the necessary ongoing performance of domestic labour nor effect obvious outcomes for its 'products'. Since some of the tasks of the housewife, such as emotional labour, are not easily contained in time and cannot therefore be made more 'efficient'[64], the question of the 'productivity' of 'improved' organisational modes becomes ever more unrelated to the exigencies of private labour. At minimum the characteristics of 'productivity' and 'efficiency' for this labour are somewhat different from those of waged labour.

It is not self-evident that, even if housewives were to work according to the time-and-motion 'efficiency' of capitalism and numerous 'home economics' educational schemes, this order in time would produce increased 'productivity' in the framework of an emotional economy or an increased rate of expropriation in sex relations of dominance/subordination. The very meaning of the category economics, whether perceived positively (in mainstream/Liberal analyses) or negatively (in Marxist and 'conflict' theories) as the 'efficient', sparing, concise regulation and use of labour and resources, must, it would seem, undergo a sea change to take account of the particularities of labour in sex relations. The contract/exchange (scarcity/competition) model cannot adequately encompass the economic dimension of women's experience. The specificity of women's labour in the private sphere and the invisible employment or only partial commodification of that specificity in the public sphere (for example, women's caring activities in offices, waged childcare and nursing) are such that these labours cannot be equivalent to 'abstract' labour in the Marxist sense[65]—a point that will be developed in the next chapter. Moreover, 'Marx's definition of productive labour in the capitalist mode of production is made from the standpoint of capital'.[66] This alone makes his analysis unworkable even for *modern* Western women's labour within the relation of the sexes because that labour is not driven or enforced solely or largely by capitalist imperatives. When it is considered that Marx also often slides between describing 'productive' labours in capitalism from the standpoint of capital and describing labour in general from this epistemological vantage point, the limits of the approach cannot be avoided.

# 2 Getting more specific: Marxism, household work studies and the particularities of women's labour

My heart belongs to Daddy . . . ? (Cole Porter[1])

Finality is not the language of politics. (Benjamin Disraeli[2])

In Chapter 1 I outlined some difficulties with Marxism's general framework and noted certain specific problems related to its account of 'the economic', such as assumptions regarding the analysis of exchange, exploitation and use of the index of time. The following discussion continues this critique, once again starting from broad issues and moving towards a further clarification of the ways in which considering women's labour suggests the limits of Marxist perspectives. In the process some aspects of household work studies are outlined which share many of the features of these limits. As Marxism and household work studies are both of importance in feminist writings on labour and economics, examination of the sexual economyths which underlie such approaches is a useful pathway towards the development of initial criteria for a feminist economics.

## Marx's definition of the labour process

Women's labour, characterised by its operation in the private sphere, is not linear. It is not a linear exchange (appropriation) with a clear temporal, spatial order and organised, enforced fixed 'prices'. There is no set or steady rate of exchange for its 'products'/services, no precise contract and no formalised system of competition/scarcity of the kind that can be distinguished in relation to wage labour. Women's labour and relations of domination/subordination connected to it occur without the timekeeping and continuous enforcing presence of an 'overseer' or a systematic administrative structure of

21

technical/bureaucratic command. That labour is largely ordered by the labourer herself around the emotional/physical demands of other people, including children, to whom emotional meanings are attached. The housewife to a large degree controls the labour process, yet her labour continues to be appropriated with or without the controlling presence of an individual appropriator to the point where it can long outlast any situation that resembles an exchange setting or contract. As Delphy notes, women continue to perform labours for their ex-husbands after divorce in the ongoing rearing of children.[3] For Delphy, aspects of what she calls the marriage work-contract continue beyond marriage, but this in itself demonstrates that Western women's labour in the relation of the sexes is quite different and organised differently from the (wage) labour depicted by Marx. The concepts and categories developed to understand the latter are thus of only limited use in analysing the former.

The problem is not simply to be resolved by describing women's private labours in terms of 'use value'. Firstly, there are some general uncertainties about 'value' theory and its application to domestic work. In this arena Folbre outlines analyses by O'Brien and Clark and Lange which assert that Marx did not perceive domestic labour as falling within the purview of analyses of 'value' since he did not judge that labour to be 'social'. Marx assumed that

> family labour could not be analyzed in the scientific terminology of 'value'. [He] treated labor itself as a nonproduced commodity—childbearing and child-rearing were considered not only unproductive of surplus value but also irrelevant to its realization. Domestic tasks were never described as aspects of a creative labor process; they were relegated to the noneconomic world of nature and instinct, analogous to a spider weaving a web or a bee building a honeycomb.[4]

Relatedly, women's private labour cannot be easily understood as 'social' or 'abstract' labour in Marxist terms—that is, it cannot be (quantitatively) equalised with other forms of labour (wage labour)—and consequently does not fall under any definition of social 'production'/'productive' labour that he employs with regard to modern societies. Domestic work cannot be straightforwardly described as labour for 'exchange' *or* 'use' in Marxist terms. Labour for Marx is not social ('value' = creating) labour or capable of conception as abstract labour in a capitalist society unless its duration, its temporality, is regulated and can be quantified:

> domestic labour does not become equal with other concrete labours and so is not expressed as abstract labour . . . [D]omestic labour does not achieve equivalence with other forms of labour qualitatively,

as abstract labour (substance of value), . . . [and] it cannot achieve equivalence quantitatively, as socially necessary labour (magnitude of value) . . . [M]agnitude of value expresses . . . the connexion that necessarily exists between a certain article and the portion of the total *labour–time* of society required to produce it' . . . Without a means to enforce socially necessary labour, domestic labour cannot be expressed in a definite magnitude of value and does not constitute . . . social labour in a commodity economy. [*emphasis added*][5]

Domestic labour in this scenario is in many ways simply outside of Marx's field of enquiry. When Marx confuses labour in capitalism with a general theory, as described in the previous chapter, the occlusion is all the more complete. While Marxist and feminist commentators frequently perceive that, within Marx's own terms, domestic labour cannot be viewed as 'abstract' labour because it is not subject to the 'general equalisation of labour'—that is, it is not comparable with wage labour,[6] the connection between this perception and difficulties in quantifying *time use* in domestic work are not the subject of discussion. The tendency in Marx's approach to reduce analysis of critical aspects of labour to time is, however, a significant theme of inquiry undertaken in this book. Moreover, even if Marx's approach to labour for 'use' were accepted, it is not evident that this approach is applicable to emotional/psychic/subjective work. Apart from difficulties in Marx's viewpoint related to encompassing the unregulated character of such work and its connected resistance to quantification, his approach seems very much associated with quantifiable 'products' and services that are visibly, 'objectively' created and consumed. In other words he concentrates upon 'products' and services which are *capable* of being produced and exchanged in commodified relations and which are not as closely tied to the imperatives of an emotional economy.

The problems associated with applying Marx's conception of domestic work as producing 'use value' are evident when one considers research in which the methodology of quantifiable comparison between women's domestic labour and market labour has actually been attempted, for example in household work studies. The equalisation/comparison of the two forms of labour has only been seen as applicable to some aspects of some domestic labours deemed broadly to be like those in the market. In other words, such studies have not assumed, as Marx appears to do, that the whole field of unpaid labour is capable of comparison. And even this more restricted comparison, as my discussion of household work studies in this and the following chapters indicates, is highly problematic and underestimates the specificity of private labour, which is unlike market labour precisely in relation to regulation and duration. Indeed, any methodology founded in equalisation/comparison must largely ignore this specificity. In this context, Marx appears to disallow specificity in his account of labour for 'use' even more than household work studies.

Marx outlines a general definition of the labour process and of 'use value' in *Capital*, Volume I, which is supposedly relevant throughout history. The labour process consists of

(1) purposive activity, that is the work itself, (2) the object on which that work is performed, and (3) the instruments of that work . . . [I]n the labour process therefore, man's activity, via the instruments of labour, effects an alteration in the object of labour which was intended from the outset. The process is extinguished in the product. The product of the process is a use value, a piece of natural material adapted to human needs by means of a change in its form.[7]

Aside from the difficulties I have noted earlier regarding 'objects' and 'instruments' and the evident inclination to perceive labour as visible, concrete labour upon inanimate, non-human materials mentioned above, Marx's framework is 'objective' in the sense that it refers to separable, discrete elements and physically evident processes or things. This 'objectivity'/externality is also perceptible in the rationalist, intentional cast of his view of labour: work is purposive and has an intended outcome from the beginning which is achieved or achievable. The spatial placement and characteristic genre of Marx's economics is indeed quite in keeping with mainstream meanings of 'the economy'. The category economics is typically not merely tied to themes of competition and scarcity but strongly associated with similarly capitalist and masculinist empirical–rational modes of thought. However one must doubt that an emotional economy is maintained by purposive, intentional labours with identified, precise outcomes. Certainly there is little room in Marx's account for less cognitive experiences and activities and for labours of *desire* that may never be fully 'known', intended, achievable or completed.

The few occasions on which Marx in this definition of general labour does allow for non-empirical processes and categories seem just as open to question. Marx refers to 'use value', which many Marxists have considered suspect, let alone feminists.[8] He also argues that the labour *process* is extinguished in the 'product', so that the labour that has gone into its creation disappears. This perception is undoubtedly very significant in describing appropriation and might also be employed by feminists. Nonetheless it is perhaps too general a statement and ignores examples that can be found in sex relations (and possibly in class relations) where the labour process may remain quite visible/physical or at least not be extinguished. For instance, women's labour process in childbirth is not extinguished in the 'product', the child. It is empirically registered on the body of the woman and on the child's body and subsequent health. The labour of child-rearing is both physically and subjectively enscribed in the body and personality of mother and child. The notion of an extinguished labour process obscures the continuing, *active* 'residues' of past labours in bodily and subjective terms that may not be a feature of labour upon inanimate or non-human objects. Inanimate objects in particular are largely given a completed form by the action of human labour. Human 'materials' do not display this finished quality.

The masking (extinguishing) of the labour process that Marx depicts as a general feature of labour in all epochs may be an element of modern Western women's domestic labour and the expropriation associated with this labour, but the disappearance of evidence of the labour process is not a priori. Moreover the masking effect, to which Marx himself contributes, derives more from the tendency to regard women's work as non-work, as unproductive. Western women's domestic labour, and versions of this that penetrate the public sphere without payment, are extinguished (rendered invisible) because they are 'naturalised' as non-work and therefore not capable of social appropriation. While Marx describes the naturalising effect of capitalist labour relations in which an historical organisation of labour comes to be viewed as inevitable, right and proper,[9] this is not quite the same process as occurs in relation to modern Western women's private work. Domestic labour is naturalised in the sense Marx outlines, yet it is also made to seem fixed or inevitable as a consequence of a literal linkage with nature/the body and of an association by extension with the 'biological' bases of personality, family, emotional life and sexual identity. Secondly, domestic labour is made to seem a result of natural properties precisely because of its emotionality. The very character of the emotional economy— labour for 'love'—assigns it voluntary, non-work status and placement beyond the sordid machinations of organised expropriation. It is no surprise in this context to find that although certain aspects of household toil (for example, 'housework') are now frequently recognised as a kind of work, if not 'real' work, other domestic labours (such as the activation of subjects and sexuality) rather more closely aligned with emotionality/love remain unacknowledged and bound to a conception of voluntarist, naturalised 'altruism'. Women's domestic labour is more thoroughly extinguished in my view by its specific connections with biology and 'love' than by a general principle of the replacement of 'product' for labour process. Marx's account of a general tendency towards the appearance of inevitable fixity in historical labour relations may be of use to feminism, but it is also necessary to point out that his analysis of this tendency in capitalism—the instance of mechanisms such as commodity fetishism in which the historicity of the labour process is extinguished behind 'fixed' commodities—reveals the differential and specific qualities of naturalised fixity as it occurs in sex relations. The domestic labour process and the expropriation of that labour process are simultaneously naturalised and extinguished in a manner which is different from and more complete than that which is evident in the waged labour process.

## Marxism, unitary analysis and singular measurement

Another element of this specificity arises with regard to the character of the labour process—that is, unlike waged labour, women's domestic labour is not singular. Whereas even the most sophisticated of waged jobs involve only a small range of activities, women's labour is defined by its conglomeration of numerous labour processes which are not necessarily highly related nor focused on a single overall goal. Indeed domestic labour and connected unpaid labours in the public sphere can be viewed as representing an entire economy rather than one job, and in this sense may be compared to the economy of capitalism. Within such a comparison women's domestic labours, though multiple, are less numerous and less precisely demarcated than the range of labours in the capitalist economy. In other words, labour within the relation of the sexes is both more diverse than a single job in the capitalist market and less diverse than the entirety of the latter economy. It exists at the intersection of the common meaning of 'job' and 'economy'.

As in the capitalist economy, there is some difficulty in defining the limits of labour attached to sex relations. For example, it is unclear whether 'voluntary' work in hospitals should be 'deemed part of the household economy or an unpaid or underpaid part of the market economy'[10] or a mixture of both. Notwithstanding the similarities in problems of demarcation, such problems are magnified in the case of sex relations because of the non-empirical quality of aspects of an emotional economy. However, if for the moment the discussion is strictly limited to domestic labour in the household/private sphere, it is clear that definitional difficulties related to the variety of women's labours are not confined to the drawing of boundary lines. Even domestic labour cannot be considered solely under a 'unitary approach'.[11] Its particular character lies precisely in the aggregation of a number of distinguishable activities often performed concurrently, and definition in terms of 'a set of features'.[12] I have already given some indication of a range of possible features describing women's private labour, within which may be included its non-unitary form. Goodnow notes that the aspect of domestic labour labelled 'housework' has itself multiple aspects and outlines some of these.

> Housework, for instance, is distinguished from other forms of work by the fact that it is done in households, usually unpaid, usually done by women, invisible, repetitive, under-valued, often undone shortly after being completed, likely to expand to fit the time available, resistant to change, and oddly difficult to pass on to others.[13]

Goodnow points out in this recognition of a complex of definitional features that any conception of women's domestic labour as a unitary construction can disguise its diverse character and its several activities. The notion of work as a single activity, she asserts, lends itself more easily to the use of single measures like time.[14] Insofar as women's work in the private sphere is multiple, it is not as amenable to time measurement, thus indicating the limits of Marx's approach to labour. Though Oakley adopts an aggregate account of the work of a housewife and makes use of quantified time, she too acknowledges the questionable value of viewing this work as singular in the way Marx does in dealing with waged labour. Oakley distinguishes the various household tasks and then adds up the time usage required for each one to arrive at a *sum* that constitutes women's domestic labour.[15] This would suggest that time may be of some use to a feminist economics, but this measurement cannot be employed in the same way as in Marxism, given the specificity of women's labour. Furthermore, when time has been used as a quantitative measure of domestic work, it seems to me that its severe restrictions become all the more transparent, demonstrating that time can only be applied to *some* extent to *some* of the several activities under the rubric of domestic work which are most like those performed in the capitalist marketplace. For example, Oakley's own distinction between labour and responsibility for labour highlights one of the areas of domestic work that are difficult to quantify in terms of time. The labour of responsibility for labour is an important element of women's experience in the private sphere, but it is not easily converted into empirically measurable units. Goodnow argues in this setting that a number of household tasks, such as children cleaning their own rooms, are designated as 'self-care' work in 'Anglo' families within Western societies, yet if these jobs are left undone by the original 'owner' they revert to the mother. Her responsibility for such tasks remains and she must therefore keep a continuous regulatory eye upon them.[16] The labour of responsibility exemplifies the timeless quality of many women's activities. They are literally unending, which cannot be said of tasks within the capitalist economy, even those involving regulatory management.

## Household work studies: exchange models and time measurement

A more obvious example of the particularities of women's private
work compared with waged labour and of the limits of time measurement occurs in studies of time usage in household labour by
writers such as Ironmonger and Sonius. They distinguish a number
of activities as household 'productive' activities and indicate that the
purpose of their research is to quantify in time these activities.
However, given the empirical orientation of the research it is not
surprising that they view household 'productive' activities in terms
of visible goods and services (including shopping and cooking) which
for the most part do have measurable time limits.[17] Even within this
framework there are some notable omissions, such as sexual activities. While they do include tasks such as childcare, which may be
seen as more closely linked to emotional labours that do not have
time limits, they appear to understand childcare once again merely
in terms of empirical, 'visible' work. A clue to this framework can
be found in their failure to include, for example, husband-care and
other emotionally charged, non-discrete tasks like the activation of
subjectivities, especially sexual subjectivities.

The diminished analysis of the character of such activities is
evident in their distinction between 'productive' and 'non-productive'
domestic work, a distinction common in household work studies.[18]
There is more than a hint here of considering women's private
labours from the standpoint of capitalist labour relations, that is
from the standpoint of the market. Ironmonger and Sonius describe
'productive' activities as those which produce 'goods and services for
members of the household'.[19] However, this restricted definition
could encompass activities they do not cover: as has already been
noted, they do not register sexual activities/services nor specifically
'husband-care'. Even within the limitations of investigating domestic
labour by employing time usage and a conception of 'productive'
domestic work, certain activities which could be included are not
mentioned. One can only point out that sexual services, for instance,
certainly appear to be an interesting omission here. These conducting
household work studies are perhaps unable to contemplate fully the
logical extension of their application of market principles to the
domestic economy. Since it would seem that sexual activities cannot
be included under the definition of 'non-productive' work employed
by Ironmonger and Sonius—described as labour that 'an individual
cannot pay someone else to do'[20]—the omission of sexual services
from discussion of measurable 'productive' household work is all the
more revealing. Indeed it suggests serious difficulties within the
market-derived analytical model employed. Either it is possible to
introduce the notion of 'productive' work into the household econ-

omy, in which case activities like the nurturing of adults in sexual and other ways may be considered measurable services or labour, or it must be acknowledged that the application of 'production'/market principles is problematic. Household work researchers appear to want to have it both ways, holding to a market-derived distinction between 'productive' and 'non-productive' labour while ignoring its implications when the result might be awkward. There is a strong suggestion here of reproducing a kind of public/private division *within* the household economy in which certain activities (particularly those linked with emotional/sexual relationships between men and women) become naturalised, outside the sphere of analysis. Nevertheless the inclusion in the mode of analysis of such activities as 'productive' services that can be measured by time usage would expose it as seriously reductive, and simultaneously undermine the conception of sexual/social 'altruism' which household work researchers depict as the underlying logic of the household economy despite their reliance on a market-oriented methodology and terms.

The 'invisibility' of much of women's work in household work studies is all the more obvious in the description of 'non-productive' work as that which 'an individual cannot pay someone else to do' (including 'sleeping, eating and exercise'). Domestic labour is split according to a measurement derived from capitalist labour relations, that is, money. 'Non-productive' activity is sloughed off as somehow not work/toil, as merely 'vital to *individual* well-being' [*emphasis added*].[21] In this account there is no acknowledgment of women's responsibility for the sleeping, eating, and exercise of others. Indeed the term 'non-productive' ignores the relationship between individual subjectivity, such as 'self-care', and women's work. Eating, for instance, is here viewed as a kind of individuated, naturalised, voluntarist activity outside the parameters of work and social organisation. This is despite the massive body of literature on connections between body, diet and women's subjectivity/activity. However, to grasp fully the parameters and significance of women's labour it would be necessary to go beyond the empirical framework employed. The unqualified acceptance of time as a measure itself produces the delimited vision of Ironmonger and Sonius's analysis. They appear to believe that time is not merely a useful instrument for investigating domestic labour but the foundation of economics per se. Against capitalism's focus on money, on *waged* labour, Ironmonger quotes Scott Burns approvingly:

> [a]nother major implication of the household economy is that money is no longer an adequate measure of our economic experience. Time, not money, is the fulcrum and measure of our experience . . . Time is absolute; money is relative . . . The hours of work done outside

the money economy rival those done inside and will soon surpass
them. Time is the ultimate unit of exchange. Money is an aberration,
an artifact of the market economy.[22]

The household work approach may be credited with taking seriously
women's labour and unpaid labour generally, and shows the polem-
ical/practical advantages of time measurement in this domain insofar
as it can demonstrate the quantifiable significance of non-waged
work. Nevertheless Burns's inclination to perceive empirical measurement as *the* mode of enquiry into economics, and time as *the* critical
measure and feature of economics—as a natural absolute rather than
an 'artifact', has the effect of disguising both the non-absolute,
historical meaning of time and its limitations in registering precisely
those labours which may epitomise the specificity of the emotional
economy. Quantifiable measurement and time are exceedingly clumsy
mechanisms for recording and analysing labours like the activation
of subjectivities which are unending, intertwined with all other
domestic (and waged) activities, do not appear to result in obvious
'products' or services, and yet may be fundamental to an emotional
economy and expropriation within it. Analysts of 'household work',
as much as Marx, often fail to comprehend fully the particularities
of labour in sex relations when they concentrate on time measurement because their framework is essentially derivative of conceptions
of labour/economics drawn from the market economy, despite their
rejection of the centrality of money. This is not only significant in
theoretical terms. The importance of household work studies is, for
example, evident in Australian federal policy initiatives around International Labour Organisation Convention 156.[23] In this context
studies by Ironmonger and Sonius and by Bittman,[24] among others,
have been used in framing federal government policy and within
related policy information materials to provide local Australian
evidence of the difficulties faced by waged workers, especially women
workers, with 'family responsibilities'.[25] Time use as measured in
these household work studies becomes the empirical method of
establishing the value of domestic work and hence carries considerable weight in providing the basis of justification for policies intended
to recognise women's labours and consequently to assist women.

However, as Meredith Edwards has argued, the framework of
time measurement cannot satisfactorily confront distinctions between
the types of work women undertake in the private sphere. She gives
particular attention to distinctions between childcare and other activities in the domestic arena. While I would add to this that a
distinction may also be made between primarily emotional
labours/emotional–physical care and more exclusively goods and
services-orientated labours, and note that all of these activities are

also invisibly present in the public sphere, Edwards's point remains vital. She notes that childcare is not as flexibly organised as other activities given that it can require continuous activity/supervision. Further, it is difficult to see technology as enhancing childcare 'productivity'. Substitution of (private) care by market purchase is more limited than for other activities, and finally, childcare competes directly with paid labour—'normally the two cannot be undertaken simultaneously'.[26] Given that different types of domestic labour appear to have special characteristics and conditions, the case for rejecting the use of a single index such as time and repudiating the exclusive focus on empirical measurement seems increasingly compelling. Goodnow in this setting outlines the potentialities of criteria of *meaning* in the analysis of domestic work and asks, for example,

> [h]ow easy or how difficult is it to move a particular piece of work from one family member to another? Which pieces of work are the easiest or the hardest to ask someone else to do? . . . Above all, what are the underlying rules or principles that make some pieces of work more moveable in some directions than others?[27]

## Towards a multiple framework and a typology of women's labours

This recognition of differences among activities and of possibly different criteria for analysis/measurement requires a kind of multiple analytical matrix for domestic labour, which may or may not be drawn together to produce Oakley's aggregate model. A framework that allows research on a variety of dimensions can then be employed to indicate diversity linked to class, for instance, and in different cultures and historical periods. The multiple/matrix approach can indeed encourage the development of specific periodisations for particular labours or for socioeconomic aspects of the relation of the sexes in particular cultures over time. Time measurement may in this setting be useful in studying the history of a strand of domestic labour, though perhaps more or less appropriate depending on the type of work under investigation, and once again not a sufficient index alone. But in order to develop an analytical matrix in which a variety of indexing criteria can be employed and which may be applicable to historical investigation, it is necessary to suggest ways of distinguishing between domestic activities/practices.

To some extent Juliet Mitchell's account of what she terms 'structures' in women's lives involves an attempt to discriminate 'types of practice'. She proposes 'four "structures"—reproduction, production, socialisation of children and sexuality—each with its own historical trajectory and generating its own oppression or transformation'.[28] The stress on specificity and uneven development is undoubtedly close to a multiple-matrix approach, but there are some problems with the distinctions she outlines. Terms like 'reproduction' and 'production', for example, are derivative of a Marxian class-based paradigm. As Rubin notes, what may be seen as 're-production' from the vantage point of this paradigm may be a 'production' in the domain of sex relations.[29] Eldholm, Harris and Young have indicated the reductionism and functionalism implied in the notion of 'reproduction' which tends to conflate the organisation of sexed social forms with 'all relations outside production relations which are required to maintain the capitalist mode'.[30] 'Reproduction' denotes an inclination to see sex relations as unproblematically essential to the reproduction of the class system, to view the former in a derivative, 'instrumental and passive manner as virtually a by-product of this external class struggle',[31] and to conceive sex relations as defined or determined by those of class.[32]

Bob Connell has advocated instead of Mitchell's 'structures' a trio of distinguishable arenas for investigating 'gender'—labour, power and cathexis.[32] These arenas, though more oriented towards structural processes than Mitchell's rather practice-based analysis, could give rise to a method of discriminating between the several types of activities involved in the economic dimension of sex relations. However, Connell's separation of power, labour and patterns of emotionality provides an undifferentiated category for labour which somehow stands apart from power and emotional investments. This structural model cannot realise domestic labours which are founded upon 'cathexis' and are imbued by relations of domination/subordination, by expropriative undertones. Connell appears to suggest that emotionality is non-economic/non-work; moreover, like Marxists, post-Marxists and many feminists, he implicitly assumes that the 'economy' is driven/defined by capitalism. 'Gender' is construed as providing a dimension to the logic of monetary accumulation which is the basis of the 'economy'. 'Gender relations' do not, it seems, have any *distinguishable* 'production' system: the possibility that 'accumulation' might mean something different within these relations is not examined.[34] No wonder emotionality is not considered under the rubric of labour.

Among models that focus upon domestic labour, an example of interest has been developed by England and Farkas. They distinguish three broad areas of practice—child–rearing, housework and '"emotional labour" (e.g., making others feel loved, secure, understood, or competent)'. England and Farkas point out that these three practices are not identical and use an index of meaning mentioned earlier to demonstrate this; that is, they assert that perhaps the task most easily shifted from one family member to another—from women to men—is child-rearing and the hardest is emotional labour.[35] In other words the activity which most epitomises the character of the modern Western emotional economy is, not surprisingly, the least amenable to change. The insight drawn from such a differentiation of labours can undoubtedly be of considerable theoretical and political value to feminists. Notwithstanding the advantages of the model proposed by England and Farkas, some further discrimination could be helpful. To this end I propose the following model of modern Western women's labours:

1 sex-differentiated *waged* work;
2 public unpaid labours which replicate features of (3) to (8);
3 'service' support labours, including travel, educational and civic duties, etc.;
4 'housework', including shopping, cooking, washing, home maintenance, gardening, etc.;
5 'body work'/body management, including organisation of diet, exercise and sleep, maintenance of 'beauty', childbirth, activities related to menstruation and health, etc;
6 sex;
7 childcare;
8 emotional labour, including 'husband-care', care for friends, neighbours, relatives, etc.

Clearly these labours overlap considerably, though some are more distinguishable than others. Nevertheless, they are not identical. The differentiation is intended to draw attention to major aspects of the several forms of work, while also recognising their tendency to merge theoretically and practically.[36] In all of these labours there is both a mode of direct activity and one of supervision/responsibility. The list is also meant to imply a movement from less personal, more public and more goods/services/'products'-oriented tasks towards personalised, private, invisible and emotionally defined activities. Thus a broad distinction between 'service' and emotionally based labours may be discovered; the former less and the latter more amenable to technological intervention, for example. The list is not precisely a continuum along these scales, for childcare may be a public and service-based activity while containing very private, emotional elements. Keeping this in mind I would add that a further scale can be perceived. While all these labours are involved in the process of 'subjectivation'—the activating of the (sexualised) self and the selves of others[37]—perhaps those that are closely linked to emotionality may be regarded as more strongly engaging both the conscious and psychic aspects of the subject and hence more linked to this process.

Such distinctions between labours and scales related to them, as well as the criteria of 'meaning' outlined earlier, can enable a greater degree of thoroughness in dealing with women's work than an analysis in which economics/'production'/labour is taken from class relations because they can reflect the specificity of that work. The themes presented here do not, of course, sufficiently depict the particularities of women's *waged* labour except insofar as these are broadly connected with the book's focus on the relation of the sexes. Further research is obviously needed to unravel the detail of the paid work undertaken by women and its interconnections with the emotional economy. Moreover, what I have discussed so far can only hint at the historical implications of a multiple analytical matrix for women's labour and how this might reveal a periodisation of Western sex relations. Donzelot, Ferguson and Hartmann's suggestions, amongst many others, regarding epochs of patriarchalism could be elaborated or rejected by employing elements of the themes mentioned above. Donzelot's propositions concerning a development from 'family' to state patriarchalism, Ferguson's similar account of the emergence of 'state patriarchy' from 'husband patriarchy', and Hartmann's depiction of sex relations moving from direct control to more indirect impersonal society-wide mechanisms, can imply an alternative periodisation from that derived through class analysis or conceptions of 'modern' society.[38] These may be assisted by accounts of women's labour/feminist economics.

On the other hand the examination of the specificity of modern Western women's labour, especially private labour, is delimited by the lack to this point of a related investigation of formal, institutional organisations, such as the state and the institutional forms of marriage, heterosexuality and motherhood. While many of these institutions have been the subject of feminist investigation[39], explicitly *institutional* analysis from the point of view of sex relations remains relatively undeveloped in feminist theory; for example there is still, despite the efforts of writers like MacKinnon and Dahlerup no thoroughgoing feminist account of the state.[40] Even if some feminists no longer view the state as a coherent entity,[41] no framework for considering the question of a feminist economics can be considered entirely adequate without more attention to institutional studies. Unfortunately, since this book cannot address all of the above issues related to women's waged labour, historical research and periodisation, or institutions, it cannot deliver a complete or final account of a feminist economics, even if that were possible. 'Finality', in any case, as the quotation introducing this chapter states, 'is not the language of politics,' nor is it the language of economics or theory.

## Beyond the market: refining a feminist perspective on economics

I have argued in this and the previous chapter that it is simply not possible to take for granted the Marxist approach to 'economics', especially when considering women's labour. Market labour can only be a metaphor for labour within sex relations and even then is probably only applicable to some aspects and types of such labour. The problems associated with economic assumptions derived from the market evident in Marxism have been further clarified by examination of similar assumptions employed in household work studies. The analysis of the latter will be continued and expanded in the following chapter, but the critical point at this juncture is the issue of the limits of Marxism for a feminist economics. While exploration of the particularities of women's labour appears to indicate that Marxian economics should be discarded as a description of *all* economic processes, this tradition can nevertheless lead to the emergence of new understandings of the character, domain, status, relational associations and causal/determining/structural elements of economics. Thus I am not suggesting that Marxism has nothing to offer to a feminist redefinition of 'materialism', history, power, politics and economics. Discussion of the limits of Marxism can indeed assist in theory development. In this setting I freely acknowledge my own debt to that tradition, though it may be of a somewhat different order from the Marxian inheritance evident in the work of many of the feminist writers I have so far considered.

On the other hand, as Harding has pointed out, what is required is 'a revolution in epistemology'.[42] A feminist economics cannot be formulated from *within* the Marxist inheritance alone. One can draw from it the beginnings of an account of the relational associations between economics and politics. Yet this cannot be undertaken without an analysis not just of Marxism but also of post-Marxism. I have indicated that Laclau and Mouffe's analysis is of limited use because of their tendency to conflate 'economics' with Marxism's account of the term and their consequent collapse of 'economics' into 'politics'.[43] The development of a more adequate model of connections between economics and politics involves a reworking of the terms. This can promote a concomitant reconsideration of structure/causality/determinacy in history that escapes the limits of both Marxism and post-Marxism. Though I reject any simple notion of either the economy or the so-called 'superstructure' (politics/religion/ideas) as *the* prime mover of history, I do not discount the possibility of broadly defining particular kinds of causalities which may be largely economic or 'political' in particular contexts while recognising the existence, effectivity and interrelation of both social processes in all historical periods. In other words a general theory of history may still be envisaged.

The position that Marxian economics cannot adequately include the specificity of women's labour means initially that if no necessary equivalence between the Marxist account and women's labour is assumed, then that specificity renders a possibly unmatched historical and political effectivity to the domain of sex relations and to a feminist economics. This may be seen as central to feminism and to its claims as an autonomous movement for social change. Moreover, if the specificity of women's labour is accepted, this implies that feminist commentaries employing Marxian categories, concepts, content or methodology must be closely examined and that those feminist analyses postulating a 'unified system' model, a synthetic socialist feminism[44], may be masking the particularities of women's experience beneath overarching unitary concepts that are supposed to cover both class and sex relations. Since these unitary concepts are often derived from Marxist theory, the attachment to the word of 'the Father'—especially in the realm of 'materialism', economics and labour—must raise a certain disquiet given the limits of Marx's work regarding women's labour.

The specificity of women's work provides yet another perspective on the problem of diversity versus unity in feminist theory. It suggests that a 'dual systems' model (at least) of economics and social life may be more appropriate to feminism. Harding states that the recognition of particularity enables the retention of categories of feminism concerning women as women, 'unstable though they be', and asserts that the inclination of an approach acknowledging specificity is towards a 'multisystems theory' as advocated by Ferguson.[45] Harding also argues that the multisystems model consists of simply settling feminist categories 'alongside the categories of the theory making of other subjugated groups'.[46] She is suggesting that a difficulty arises here in the extent to which the model tends to leave other categories untouched. However, as I have already indicated, close consideration of the specificity of women's labour can give rise at minimum to some evocative hints with regard to Marx's theory of labour in class relations, for example, that might alter Marxian categories.

Harding goes on to note that a multisystems approach 'leaves bifurcated (and perhaps even more finely divided) the identities of all except ruling class white Western women'. But this concern surely rests on her assumption that other theories remain unchanged and unconnected with sex relations by the maintenance of distinct feminist categories. There may be some danger of the splitting and/or dispersal she describes. On the other hand, bifurcation may be reassessed as not merely negative politically and theoretically—she depicts bifurcation as leading to a 'fundamental incoherence'—but as a resource and a basis (even touchstone) for theory development.[47] Harding finally postulates that the multisystems model involves relinquishing 'the totalizing "master theory" character' of feminist analysis. Though I do not necessarily object to 'totalizing' elements in theoretical endeavours per se,[48] it may be as well to abandon a notion of feminism as providing *the* master theory for all social relations in all cultures throughout history.

Harding suggests that 'another solution' to the issue of diversity versus unity and the specificity of women's experience may lie not in a multisystems model (which conceives attributes of women's experiences as both shared with and different from those of other social groups) but under the banner of strategic 'solidarity around those goals that can be shared'.[49] This viewpoint has something in common with Laclau and Mouffe's valuation of politics as central to theorising. Yet it goes beyond this, for '[f]rom this perspective, each standpoint epistemology—feminist, Third World, gay, working class—names the historical conditions producing the political and conceptual oppositions to be overcome but does not thereby generate universal concepts and political goals'.[50] What Harding does not mention in her account of the solidarity model is that each standpoint epistemology may continue to retain certain broadly 'universal' concepts and goals *within* the arena of its own theory and that the notion of solidarity may not necessarily be at odds with a multi-systems approach. In keeping with Harding's own framework regarding the instability of feminist categories, it is possible to have a feminist epistemology/politics that can make use of the uncertain relationship between a multisystemic paradigm of diversity and a strategic solidarity that is not to be confused with singular political organisation or concerns. The conception of the specificity of women's labour and a feminist economics derived from it can enhance and draw upon this unstable combination. It is in the light of these considerations that an examination of attempts to conceive 'materialist' analyses of women's position may be undertaken. I have indicated Laclau and Mouffe's failure to depart sufficiently from Marx's 'materialism' in significant ways. They seem haunted by the presence of their 'father'. Having determined the limited vision of the paternal perspective, it is necessary to investigate further the degree to which household work studies and feminist 'materialisms' elaborate 'a revolution in epistemology' or to which they too may remain bound by the economic assumptions of the 'Daddy' of political economy.

# 3     Household work studies: closer to home?

Political economy: the Dismal Science. (Thomas Carlyle, 'The present time', *Latter Day Pamphlets*, 1[1])

From women's eyes this doctrine I derive:
They sparkle still the right Promethean fire;
They are the books, the arts, the academes
That show, contain, and nourish all the world. (William Shakespeare, *Love's Labour's Lost*[2])

## The system of the 'household economy'

In the preceding chapters I spent some time enunciating the ways in which modern Western women's unpaid labours, especially in the private sphere, might be considered to display a specific form that differs from the forms of labour undertaken in the public sphere of waged work. This analysis not only suggests particularities of labour in the private as against the public realm but a superimposed framework which differentiates labours performed by Western women from those which Western men undertake. While women engage in private and waged work in modern Western societies, they tend to be defined by the first and are uniquely active in that arena compared with men. The specificity of women's labour is therefore epitomised by, though not exclusive to, the characteristics of the private domain. I have consequently argued that the *private* labour of Western women demonstrates the form of an emotional economy that crosses the private–public divide and is critical to the relation of the sexes, that is, it is defined by a linkage to the organisation of power in sex relations embedded throughout the private and public spheres of Western societies. In postulating a specific economics associated with the private sphere and indeed a separate economic 'system', I am in many ways merely reiterating a view put forward by various analysts of 'household' work and the 'household economy'. Duncan Ironmonger, whose analysis is not strongly informed by feminist perspectives,[3] is nevertheless able to assert that the

household sector is large enough and distinctive enough in its
method of operation to deserve the term 'household economy'. The
other sectors of the economy can then be called the 'market
economy' . . . [T]he transactions between household and market are
perhaps more akin to international trade between two countries than
transactions between different industrial sectors of the single
economy.[4]

Ironmonger recognises here that household 'production',
'consumption', and labour time use, for example, are not just
dimensions of the 'market' economy.[5] He values the contribution of
the Australian household economy as three times that of the manu-
facturing sector or ten times that of the mining sector.[6] This economy
has also been valued at between $90 billion and $146 billion a year.[7]
Michael Bittman, using Australian Bureau of Statistics 1990 figures,
estimates the total value of unpaid household work at between 52
per cent and 62 per cent of GDP.[8] He found that women overall do
70 per cent of the work in the household economy (regardless of
traditional class variables such as male income), while Ironmonger
states 75 per cent of all time put into 'household productive activities'
is contributed by women.[9] The scope and particularity of the economics of the household sector justify its claims to being a distinct
form. Though Ironmonger's use of the term household economy is
founded on Australian empirical data, there appears to be a strong
basis for acknowledgment of such an economy in other Western and
non-Western nations. Household work 'was estimated to be at least
40 per cent of the GNP for the United States' in 1980 and 'was
probably more in less monetarized economies'.[10] These estimations
are likely to rely on the most limited account of what constitutes
household work—that is, the 'production' of empirically evident,
concrete goods and services that are already measured or are capable
of being measured by the indices of the market economy. The scope
of household economies has undoubtedly been grossly underestimated. However, even accepting the limits of these estimates for
nation states, the vast field of household work in various cultures
supports a conception of an economic domain outside the market
economy. The contribution of this labour is regularly not assigned a
value to 'the economy', is not present in official statistics and is
rarely included in national income accounting.[11] Furthermore, there
are no established international guidelines for measuring or conceptualising unpaid household labour.[12]

The writings of household work analysts like Ironmonger could
certainly be employed in a feminist economics in that they acknowledge an economics which is imbued with non-market assumptions
and practices concerning labour, constituted largely by the labours
of women,[13] of a considerable size (and yet frequently invisible in

theoretical, empirical and policy terms), situated in a specific domain
(the household sector) and asserted to be in competition with the
market economy for labour and in relation to 'time-off' ('leisure'
time).[14] In these ways the framework of household work may
contribute to feminist perspectives which call for women's labour to
be seen as *work*. The employment of household work approaches in
conjunction with feminist analyses enables researchers to ask questions about women's private labour and conceive its social
significance in similar ways to approaches dealing with waged
work.[15] This incorporation of women's labour into labour and
economic studies corroborates and promotes a place for a feminist
economics, a feminist 'materialism'. Additionally, it might allow for
recognition of the worldwide and cross-cultural existence of a field
of economics linked to women's labour that can alter understanding
of 'modernity' and Third World 'development' in contemporary
societies. Such an incorporation might also encourage research on
possible historical continuities, in other words supporting the notion
types of non-market economic systems over time.

Ironmonger grants the arena of household work the status of an
economic 'system' in contemporary Australia. This appears to provide some basis for accepting Giddens' usage of the term 'system',
thus allowing for the application of an historical dimension to this
area of labour. Marxism within such a framework has no exclusive
or privileged claims to ownership/employment of historical materialist analysis. Giddens describes 'social systems' as a terminology 'in
which structure is recursively implicated' and comprises 'the situated
activities of *human agents*, reproduced across *time* and *space*'.[16] The
continuity of a process of structuration involving the combined
constitution of structures and agents may well be a useful means of
viewing women's labour within societies, cross-culturally, and
through history. The notion of 'economic system' could therefore
offer a variety of possibilities even if perceptions of time and space
in Giddens's work are not always sufficient or appropriate to feminism.

The writings of household work analysts can in this usage present
a specificity within women's labour whose very distinctiveness indicates that its character is not simply arbitrary, or entirely temporary,
disorganised or fragmentary. Specificity refers to the presence of a
*logic* of operation. Clearly the conception of a specific logic, a
'system' of economics, does not necessarily imply an unaltered
continuity but merely the possibility of an organisational form which
may vary in particular social contexts. The variations may or may
not exist over long periods. Rosaldo has proposed that 'male dominance, though apparently universal, does not in actual behavioural
terms assume a universal content, or a universal shape'.[17] Whether

or not hierarchy is indeed 'universal' (ubiquitous) in sex relations, I suggest in this context that hierarchical sexual orders can be conceived as possessing an economic organisation which is an aspect of the content or shape of these orders. I allow that the economic organisation of hierarchical sexual orders may have continuous ('universal') features.[18] Nevertheless it is likely to take on particular forms in particular contexts, that is, to be constituted as particular economies or 'economic systems' framed within limited periods and cultural contexts.

In proposing a conception of a distinct 'economic system' of household work in Australia, Ironmonger gives credence and status to a feminist epistemology of power in sex relations in modern Western societies. The 'market' economy cannot be seen as the be-all and end-all of economic studies in the light of such work. Moreover, by implication, analysis of the specificity of Western household economies can raise questions regarding the economies of non-Western sexual orders and those of previous historical periods and societies. Household work provides a potential language for referring both to a general economic organisational form and to particular economic 'systems' connected to the myriad of hierarchical sexual orders, that is, it can be employed to envisage a feminist historical materialism. In these senses household work analysts offer theoretical insights and empirical information which could be advantageous to feminism and perhaps invaluable to a feminist economics.

## Household work studies and the 'unit' of analysis

Nevertheless, the analyses of unpaid household work undertaken by commentators like Ironmonger and Bittman and by organisations like the Australian Bureau of Statistics, as well as studies undertaken in other Western countries,[19] contain some serious problems for feminist accounts of women's labour. Such analyses are explicitly delimited and implicitly restricted by a series of assumptions which may be problematic from a feminist perspective. The specificity of the economic organisational form is largely defined by and confined to a spatial *domain* or *site* articulated in terms of its negative relationship to the market economy: it is described as encapsulated by the boundaries of the 'household' unit and as that which is *not* the market economy. This may present historical and cross-cultural difficulties for feminist approaches which attempt to investigate women's unpaid/domestic labours in societies which do not distinguish public from private–domestic in the modern Western sense, or which have little or no market economic systems. The presumption of a discernible binary, a sharp demarcation, constituted by a dis-

tinction between household economy and the economic organisation
of everything else (the market economy) could be said to be appropriate to modern Western societies and therefore a useful framework
at least within the parameters of these societies. Nonetheless, the
presumption contains a number of problematic features which remain
even if it is only applied to the limited cultural and temporal range
of 'modern' Western social relations.

The household work framework can indicate, by emphasising the
domain category of the domicile/household, the generally
unrecognised considerable *extent* of unpaid/domestic economic activities and their significant *contribution* to the total of economic
processes, suggest the particular *character* of *non-market* work, and
bring to light both an arena of *women's work* and the *greater
contribution of women* within the household. But the concentration
on domain directs the commentaries towards a supposedly neutral
demarcation which takes as given a site, as if it had a self-evident
status and walls like those of a building: the category of 'household'
indeed encourages an image of a building (the home). The method
of analysis proceeds from the starting point of the 'factory building',
not from the power relations that might surround, imbue and in fact
construct the meaning, activities and image of the building. This
corresponds with a kind of depoliticised neoclassical economics (of
the market) in which the descriptive category of household 'sector'
or 'unit' is the primary signifier and sex/sexual hierarchy is thereafter
discovered as a feature of it. Household work is given the status of
an economic or 'production' system which is then, by implication,
almost incidentally conceded as revealing attributes of sexual hierarchy. But if domain/site rather than (hierarchical) sex relations is
viewed as *the* defining principle of women's domestic labours, (a)
the character and (b) the extent of the specificity of these labours
will be conceived in a certain way and, from a feminist perspective,
in a limited fashion.

## Introducing power: reconceiving the 'familial' in terms of a political economy

In relation to the character of women's labour, it is immediately
evident that the notion of the 'household economy' is not a framework of political economy, though questions of power *may* be
examined. Political economy takes power as endogenous to labour
and economic organisation and conceives institutional/sectoral/'unit'
analysis as a useful but partial and potentially dangerous approach
insofar as the latter can simply fail to recognise, or be employed to
disguise, relations of domination/subordination by taking for granted

commonly accepted and therefore probably hegemonic understandings of social life. I have no doubt that there is a very important place for institutional/sectoral/'unit' accounts in economics generally and in relation to households and women's work. Moreover it is clear that such accounts can provide empirical data in particular and some conceptual questions for a political economy framework. But household work debates often do not evince awareness of the theoretical issues and methodological assumptions employed, which may cause the work to reflect covertly and overtly particular political stances rather than the neutrality of categories and concepts that is implicitly claimed. This has implications for assessments of studies of household work as a form of *feminist* research. A political economy perspective by contrast enunciates its politics and takes oppression as its starting point.[20] This standpoint framework can be called upon to answer for its domain assumptions,[21] and can incorporate theoretical paradigms as varied as post-structuralism, certain versions of pluralism and liberalism, and Marxism/post-Marxism. Furthermore, it can actively draw upon the possibilities of various feminist positions and centre upon the context of sex relations.

The limitations of a 'household' work approach with its comparative de-emphasis of power and sexual hierarchy may also be shown in the slippage from a building metaphor to a familiar one. The building aggregate is confounded with a 'family' aggregate which bears a similar tendency to ignore power and historical or cross-cultural specificity. The 'family' aggregate, like that of the 'household', is descriptive. It includes the labours of all the people in the 'family' unit who are seen as contributing more or less to a cohesive domestic economy. The inclination of institutional/sectoral/'unit' approaches to underestimate or ignore differences in the meaning and character of the labour of women, men and children beyond acknowledgment of different amounts of work performed or time taken to perform it, and the connected disposition to perceive 'the family' as an undifferentiated unit, have been criticised by feminist writers in theoretical, empirical and historical terms.[22] Apart from problems associated with the overly general term 'the family', the most obvious flaw in treating familial forms as units is evident when this leads to dealing with 'the family' as if it could be represented by or is synonymous with the (male) head of household. The implication of a household or family framework is that analysis can proceed from a notion of cohesion and commonly shared interests and that differences between members of the 'unit' under study are not seriously divisive. The approach bears a resemblance to liberal and liberal-pluralist models of social relations and is by no means conceptually neutral.

This is certainly one way of perceiving women's labour but it is not politically innocent and has the limitation of underplaying precisely what might be of interest to feminism, that is, it tends to deny a separate status to women and can eclipse the benefits which accrue to men from women's labour. Such an approach gives rise to recognisable methodological consequences and conceptual closures. It is no surprise, then, that Ironmonger's research, along with many other household work studies including research by the Australian Bureau of Statistics (ABS), does not acknowledge labours like sexuality or husband-care. Ostensibly this appears to be because certain labours within the household are somehow not considered to be 'productive' or 'economic'[23] even though, as I have noted, it is by no means evident why this should be so. However I do not consider it is coincidental that Ironmonger and the ABS list a variety of labours described as care, but no activity, let alone care, is identified as *for* men. The forms of care noted are all able to be viewed as labours *for* the children or *for* 'the family'/'the community'. The recognition of any work as simultaneously by women and for men might fracture the implicit conception of marriage/family/household as built upon equal exchange and sharing commonly found in household work studies. This conception can exist side by side with a recognition of generalised inequality related to women's greater contribution to household labour.) The political connotations of husband-care, which suggest more specific benefits arising from this inequality, perhaps partially explain its absence from these studies and reveal possible circumscription within the household/family framework.

Like Delphy, I therefore see some advantages in a vantage point which takes power as a central feature of examinations of labour and sees the 'household economy' outlined by Ironmonger as more usefully investigated from a feminist perspective in the explicit terms of political economy. A focus on power may alter the priority given to the institution of household/family or at least clarify its meaning in connection with forms of social hierarchy—in particular sex relations in modern Western societies. (I will outline possible alternative institutional methodologies in a later chapter.) Additionally, a political economy focus enables not merely analysis of inequity and appropriation but can trace alterations (intensification/reduction) in power relations. To the extent that studies of 'household' work and the 'household economy' may underestimate differences in power, they are less capable of locating change. I concur with Elshtain that unless social theory offers potential for change, it is empty.[24] This suggests further problems in the use of the notion of household economy and indicates the benefits of a *located* analysis which specifies power as its central concern. By this means accounts of economic processes hopefully escape Thomas Carlyle's estimation of political economy as 'the Dismal Science'. Indeed, the household economy approach of Ironmonger and the ABS is problematic for feminism because it retains more that is both dismal (descriptive and comparatively static) and 'scientific' (empirical and quantitative rather than meaning-oriented) than the political economy orientation exemplified by writers like Delphy.

This is not to suggest that Delphy's schema should not also be subject to investigation. Her approach is analysed in detail in Chapters 6 and 7. But Delphy's work is illustrative of a type of critique which, I suggest, offers greater possibilities to a feminist economics. Delphy's mode of examination is, after all, an explicitly feminist one: household work studies are not always or necessarily so. The aspect of Delphy's feminism that is, in my view, critical to revealing elements of the theoretical *and* empirical particularities of women's labour is its prioritising of power. Household work studies are often limited to the degree that they are ambiguous about, de-emphasise, or do not prioritise power. The epistemological framework for a 'materialist' feminism, a feminist economics, cannot decentre power without cost, for prioritising power enables a more adequate recognition of specificity. Household work studies fail to perceive the full extent of this specificity, which, even in their own empirical terms, exposes constraints in their analyses insofar as they are unable to recognise certain important conceptual, methodological and concrete elements of modern Western women's private labour.

## Household work studies and the privileging of the market

Household approaches usually revolve around an epistemology describing women's private labour as non-market work. It is a definition largely derived from the market. Women's private labour is conceived as a negation, as that which is not. The standpoint of capitalist labour relations is a method for analysing such labour, yet it does not go very far towards consideration of the possibility of any other *logic* for economics. This is cross-culturally and historically restrictive. In addition it fails to perceive the potential difficulties of the hegemony of the 'exchange' model in labour studies pointed out by Hartsock in relation to both non-Marxist and Marxist research.[25] A 'market' standpoint involves a set of assumptions that are not argued, merely asserted, and can lead to difficulties of the kind mentioned earlier in connection with conceptions of the undifferentiated 'household'. Because the market standpoint proposes an essentially derivative analysis—the logic of household labour is that it is *not* market—it can slide into treating household economics from

46                                              SEXUAL ECONOMYTHS

the class/'stratification' position of the male wage earner, the supposed representative of the household. When the framework implies
that capitalism (the market) and class practices are *the* measure of
all social relations, there is an obvious inclination to suggest that
class *creates* social life, despite the resultant effect of eclipsing the
impact of sex relations and inequitable benefits to men. If class is
the measure of and creates sociality, it becomes possible to see
class/'stratification' as providing the ranking of the social totality, in
which case male heads of household can stand in for 'the family'
and represent women and women's labour.

To construe household work as non–market labour seems at first
a quite reasonable starting point, but the logical extension of the
market standpoint is to deny the specificity of sex relations and
sexual hierarchy, let alone the specificity of women's private labour.
If one agrees with Garnsey and Delphy that class cannot produce a
ranking for all social relations,[26] then the initial starting point must
also be reconsidered as derivative and reductive. The disposition to
view all labour as related to the capitalist market reflects a failure
to imagine whether labour and economics could be conceived in
different terms, and whether there would be any advantage in a
different perspective, rather than any considered judgment of alternatives. The market standpoint is closely aligned with the domain/site
focus of household work studies. These approaches are often combined, and both show signs of class-based and masculinist
assumptions in that they presume a priority for the sphere of waged
labour which is associated with public activities and with men. This
priority can occur explicitly in household work studies despite a
supposed concern with the specificity of the household sector, or
implicitly even when the connections between the market and household economies are said to be at the level of hypothesis and are
suggested as indicating competition. In both cases the private sphere
and indeed sex relations are construed largely as passive adjuncts to
the public marketplace. Jamrozik, for instance, overtly defines the
household economy as a dimension of the market:

> the household economy does not function separately from the market
> economy but to a varied degree it forms a part of the latter, serving
> mainly as a field for the consumption of goods and services
> produced in the market. Secondly, the degree of autonomy a
> household can achieve is related to the social class of its members
> . . . As a result, class inequalities in the household economy reflect
> rather than countervail the class inequalities of the market economy.[27]

Even if it were accepted that the household and market economies
are largely intertwined, Jamrozik goes beyond this to assert that they
are one and that one is capitalism. There is no suggestion here that

the market economy forms part of the household economy or that sexual inequalities in the market economy reflect the sexual inequalities of the household economy. Jamrozik might not disagree with the latter statement, though probably he would disagree with the first, but so one-sided is the analysis that neither occurs to him. The commentary is always directed towards demonstrating that the market *drives* the household. Ironmonger, who speculates that the market competes with the household economy, might appear to ascribe some effectivity and active dynamic to the latter. However it seems the hypothesised interaction/competition remains at the behest of capital: 'we have the general hypothesis that . . . [w]hen the market is booming with a strong demand for labour, the household will do less work at home. When the market is recessed the household will do more work.'[28]

While, as can also be said of Jamrozik's work, Ironmonger's assertions may be included in a feminist economics and are certainly not inimical to it, what is of concern is the failure to pay equivalent attention to the demands of the household economy. These writers advocate a one-way flow that is class-based *and* masculinist to the degree that the market appears to control everything and sex relations are rendered relatively invisible. The specificity of women's labour and its organisational forms are driven underground in an overarching paradigm which in the end conceives the household (and sex relations) as mere dimensions of the market. Ironmonger's view of 'competition' between the two economies is undermined by his failure (like Jamrozik's) to note any disjunctions or disharmony between them. They do not in fact compete, for in his analysis the household only ever picks up whatever labour is left over. Moreover, the notion of 'competition' as the sole terminology to describe connections between these economic systems is itself based in a market model which may not adequately take account of other interactions that might occur, such as different forms of accommodation by either or both systems.

The point here is that no matter whether they argue that the household economy is part of the market economy or in some (limited) sense in competition with it, the household unit/family and market standpoint frameworks of many studies attending to women's private labour reduce the specificity of that labour by derivative and pacified analyses of the household economy. The conception of the household economy as a largely passive adjunct to the market is upheld despite evidence, of which these same commentators are aware, that connections between household and market economies are weak:[29] women's responsibility for domestic labour is little affected by their market labour and relatedly men's contribution to household work appears to remain the same even if their wives take on paid work, if a child arrives, or if their own paid work is reduced.[30] Change in the sphere of paid work is not associated with 'anything like equivalent change' in household labour.[31] While accounts of connections between household and market economies are likely to be of use to a feminist economics, what is of concern is the advocacy of a one–way flow between the two which constitutes the former as an appendage of the latter in the face of evidence to the contrary. Such household studies are frequently not merely somewhat depoliticised in terms of not providing a political economy focusing upon power relations but, to the extent that they acknowledge power, that power is often seen as ultimately lying in the public (male) world of capitalism/the market. Paradoxically, household work analyses may contribute to the theoretical invisibility of women's private labour.

## The problems of market-derived definitions and valuation measures

These analyses may implicitly or explicitly acknowledge the limitations of the concept of the household/family unit but much less often display awareness of feminist debates over the connections (or disjunctions) between class and sex 'systems' that are directly related to the assumptions of a market model. The latter problem is even evident in examinations of household work that offer a feminist perspective. Yet the consequences of using a market model are concrete. For example, as noted in the previous chapter in the context of Ironmonger's research, if the orientation is exclusively towards market-based measures, then only certain kinds of labours will be conceded the status of 'productive' or 'economic' activities. This narrowed perspective is common in household work research. For example, Canadian and American studies reveal a similar circumscription in their definitions of household labour.

> The value of those economic services produced in the household and outside the market, but which could be produced by a third person hired on the market without changing their utility to the members of the household.[32]
> Household work is defined as non-market uses of household time that result in the production of a good or service that could be purchased in the market.[33]

Despite the simultaneous recognition of a specific and differential logic to household work, this particularity is reduced to a negative adjunct of the market which can only be investigated in market terms. The minimum effect of such an approach is to ignore a range of activities that might be deemed to epitomise the emotional economy, including emotional labours like the unending 'production' of sexual identity that cannot easily be translated into a waged job undertaken by any 'third' person or a purchasable service. Secondly, the market model requires estimations of household labour as 'valued' by the market. While household work studies analyse the domestic economy according to measures derived from the market, which is evident both in the use of the index of time and the concern with those labours which are deemed marketable,[34] these studies also intend to provide a comparison with the market. Consequently, the household economy, though initially measured by dissected time, is ultimately valued in thoroughly market terms—that is, money. In this context the United Nations *World Survey on the Role of Women in Development* (1989) recognised the continuing pressure to include unpaid work in economic statistics and planning but also noted that this work had 'to be valued in the monetary terms prevailing in the market'.[35] Methods employed in household work studies are in keeping with this viewpoint. The methods generally employed are (a) the market cost/individual function replacement cost approach, summing the cost of separate services usually discerned on the basis of low-valued 'skills', (b) the housekeeper replacement cost approach and (c) the opportunity cost approach, estimating the potential earnings of or earnings forgone by the person devoting time to unpaid household work, that is the 'housekeeper'. Studies from Australia, Canada, Finland and France, for instance, make use of these methods despite the fact that putting a money value on women's private labour transposes a conception of 'rationality' and cost effectiveness from neoclassical economics to an arena that is clearly understood as driven by a different logic.[36]

All these writings recognise the particularities of the household economy insofar as the unpaid work performed is crucially linked to a specific group of workers—women—and is driven by a non-commodified logic. Furthermore, the market valuation methods applied to that work are indirect and therefore rather imprecise and experimental. Nevertheless there is virtually no discussion of the assumptions behind the derivative paradigm which estimates the 'value' of one form of labour/economy from the standpoint of a different kind of labour/economy. The difficulties of a derivative model are perceived largely as simply affecting the accuracy of the research and not as implying the need for a fundamental reconsideration of the epistemological and methodological framework.[37] The advantage of the market model is, of course, that women's private labour can be considered in national income accounting and brought into the economic code of the public arena. Moreover, within this, different methods can be employed for particular political and economic purposes. Meredith Edwards appears to see no problems with the market schema but notes that if the concern is to take 'the point of view of the home manager', in other words the vantage point of women, the 'opportunity cost' approach may be the most appropriate.[38]

Whether or not this is the case, and without minimising the advantages of including women's private labour in mainstream economics, it is also necessary to reflect on the implications and possible disadvantages of a model which presumes that this is the priority, given the links between such a viewpoint and Liberal/orthodox Marxist perspectives which claim that feminism should be directed by an intention to enter the public male world. Such a claim is contentious. As Elshtain and Jaggar have asserted, substantial doubt exists as to whether the lives of women can or should be measured according to the values and practices of the public and capitalist domain of modern Western societies, and whether such a stance is likely to lead to the feminist aim of the abolition of sexual hierarchy.[39] In the light of these criticisms of epistemologies which propose that women, women's activities and the private sphere should be largely brought into the public arena—essentially that women should be treated as men and become like them—the market model is one-sided and has a particular political context which is potentially restrictive and even strategically dangerous.

### Rethinking sexual economyths: challenging market epistemology

It is important to take seriously Edwards's point regarding the use of different methods (and concepts) for different purposes and apply this beyond her own market-based approach. If a feminist economics intends to concentrate upon the power relations of sex and/or the specificity of women's labour in class, race, and sexual hierarchies, household-unit based accounts that investigate labour from a market standpoint are of limited use. This is especially so since such accounts do not even inquire whether the acknowledged differential logic of the household economy might be measured or examined *directly* rather than indirectly, and ignore the question of whether the market economy could be conceived from the standpoint of the household. Household unit/market model analyses always assume that conceptual development and change and therefore, implicitly, political change should be in one direction. Women's private labour can be conceived in public capitalist (male) terms but apparently never the other way around: the subject of the market paradigm never becomes the object of the other's gaze. Indeed, there are strongly masculinist elements in the paradigm which have consequences for its ability to produce a feminist analysis. Campioni and Gross's description of phallocentric epistemologies is relevant to this debate: the 'objects' of knowledge, women,

> do not . . . produce knowledge themselves . . . They can be known, but not themselves knowers . . . Knowledge mediates between the subject and the object in one direction only . . . The opposition between masculine and feminine constitutes a picture of . . . social relations which, while it gives a place to women, does so only in terms of what they are not and never can become: independent categories of analysis. Femininity is seen as deficient, variant masculinity. Women can only exist dependently. There can never be any question of a theoretical perspective which takes women as subjects, nor takes its point of departure from women's specificity.[40]

Relatedly, Victor Seidler claims that an inclination to transform demands requiring recognition of women's unpaid labours into an externalised, instrumental account of a measurable number of hours corresponds to a particular version of Western masculinity.

> It is easy for us (men) to turn everything into a task. . . In this way our experience becomes externalised. . . This form of instrumentalism has a deep connection with masculinity. It makes us feel capable in the public sphere. It is more difficult when the personal is claimed by feminism to be 'political'. This is something we can often only cope with if the personal itself is redefined as a series of externalised demands and obligations involving so many hours of housecare and childcare.[41]

This suggests that household work studies (and Marxism), given their concern with the empirical measurement of time and their public arena—market standpoint, may be bound by a phallocentric perspective in a further sense. If the dominant conceptual framework for investigating women's private labour in modern Western societies is phallocentric, it is little wonder that the framework cannot adequately deal with sex relations, sexual hierarchy and specificity of women's position and work. I have indicated previously the usefulness of a multiple-matrix approach to women's labour and particularly their private labour. Nonetheless, the masculinist features of household work/market model studies suggest that although these approaches may provide some kinds of material for a feminist economics or could be critically employed in certain circumstances around political projects concerned with equality/sameness in the public sphere, and perhaps when dealing with some aspects of class differentiation between women/'families', they are not adequate for a feminist political economy. The hegemonic status of the household/market approach may in fact continue to mask the conceptual field and act as a counter to such a political economy. On both grounds there is a strong case for the requirement of alternatives to household work/market accounts in which women and their activities are not defined by covering terms like 'the family'/'household' or by what women are not or do not do.

The usefulness of a feminist political economy which does not view women in dependent/negative/deficient terms, does not rely upon the 'Fathers' of Liberal/Marxist economics, and takes women and power in sex relations as the *subjects* of the analysis, can scarcely be overstated. To conceive women and sex relations as a priority is to attempt the development of new and 'independent' categories of analysis in the sense of, at minimum, foregrounding a non-derivative conceptual framework employing *direct* methods of investigation. This is not to propose modern Western women's experience as sufficient for a theory of feminist 'materialism' or feminist economics, since power in and the organisation of sex relations in Western societies goes beyond this. It does on the other hand involve the constitution of women as agents whose specificity may reveal something of relevance for a variety of conceptions of generality, universality, totality and historical continuity and change, rather than merely suggesting the idiosyncratic particularities of epistemological and empirical specialism. As the quotation from *Love's Labour's Lost* heading this chapter asserts, women (and sex relations) can invoke an academy as various as that which covertly or overtly perceives men as the subject of intellectual speculation. The book's focus upon the question of a feminist 'materialism' and a feminist economics, especially with regard to the modern Western private sphere, is therefore not of marginal or exotic interest but pertains to the 'mainstream' of philosophical and economic inquiry. Specificity does not in this instance lead only to a specific vantage point.

However, to undertake an analysis which constitutes women and sex relations as a priority does appear to demand—as can be seen from the critiques of Marxism and household work studies, for example—a significant reconsideration of epistemology in order to evade or at least ameliorate the hegemony of derivative and phallocentric perspectives. I have touched upon some of the difficulties of remaining bound to paternal epistemologies like Marxism and Liberalism and the limits of these frameworks in relation to examination of the specificity of women's labour. The issue of the extent to which the word of the 'Father' is reconceived, and consequently to which specificity can be investigated, is raised once again in explicitly feminist accounts of the arena of women's work. Having discussed some procedures for developing a feminist political economy, it is now appropriate to look more closely at the work of those feminist writers who have concentrated upon women's labour, that is the work of so-called 'materialist' feminists, to see how they deal with these problems. Do they remain trapped, in their case, by the

Marxian paradigm and are they able to deal adequately with specificity? Do they in fact produce a *feminist* 'materialism' which privileges women and sex and therefore challenges the hegemony of household work/market model conceptions of unpaid private labour? I debate these points by reflecting upon 'materialist' feminist analyses of the connections between the sex and class 'systems' and the 'household' and market economies.

# Part II

# 'Materialist' feminisms and a sexual epistemology of economics

# 4

## Towards a 'materialist' feminism?: unity or multiplicity in 'the Dismal Science'

With a name like yours, you might be any shape, almost.
(Lewis Carroll, *Through the Looking-Glass*[1])

I have argued that household work studies tend to underestimate (a) the *character* and (b) the *extent* of the specificity of women's labour, in the first instance insofar as they underplay the significance of power and in the second, as a result of their derivative definition of the household sector in the terms of the market as negative and passive. In some ways, then, these studies are contradictory for they acknowledge that the household economy is different from the market (capitalist) economy, yet retain a perspective which to a considerable extent ignores this difference. Often, in descriptions of what constitutes 'economic' labour and in measures of the 'value' of that labour, they conceive the household economy as possessing no specific ('independent') power relations or dynamic. This ambiguity is paralleled by debates amongst feminist 'materialists' about whether sex relations—especially those in the private sphere of modern Western societies—do have a separate, distinct economic 'system', whether the economics of women's labour in the private arena are derived from the class 'system' or whether labours performed in the public and private domains may be seen as inextricably interconnected and therefore best analysed by a unified model that spans both. The orthodox Marxist position which refuses any separate status to sex relations and a sexual economics, and perceives women's labour as an aspect of capitalist relations, can be said to have been largely discredited. While sex relations—or rather sexual hierarchy (patriarchy)—are now generally conceded some distinctiveness from capitalism, the *extent* of specificity remains an issue, as it is in household work studies. There have been two major responses to

57

this problem which have attempted to overcome the perceived limits of orthodox Marxism at the same time as retaining a concern with 'materialism', economics and labour.[2] The two responses correspond to different hypotheses about the connection between household and market economies in household work research.

Some feminist 'materialists' have argued that the sex and class 'systems' cannot be studied separately, despite their particularities; that is, these writers have adopted a 'unified system' model. By contrast, another group of feminists have outlined a 'dual systems' or 'multisystems' approach which asserts that sex and class (and race) are distinct categories and must be analysed separately before any discussion of connections or disjunctions can be undertaken. The latter framework appears to provide greater recognition of the *specificity* of women's labour, which I have suggested is critical to a feminist political economy, but closer examination of these accounts is necessary to assess what they offer and how far they have gone in reassessing the conception of labour put forward by Marxism.

## Unified system models: some questions

Examples of a 'unified system' model occur in the work of Vogel, Young and Jaggar. However there are significant variations among these accounts. Vogel's account of women's labour involves comparatively little reconception of orthodox Marxism and tends to interpret women's labour through the lens of class analysis, which raises once again the serious limitations of an almost exclusively derivative paradigm. Vogel's work assumes that there are no disjunctions between the organisation of women's unpaid labour and the waged labour system.[3]

In this context, Hartmann's analysis implies that a straightforward functional connection between sex and class hierarchies may be strongly doubted. For example, she notes that during periods of economic expansion working men's resistance to women's labour may not function in capital's interests, and that capitalism's tendency to reduce differences among workers may threaten patriarchal relations.[4] This suggests that the relation of the sexes and class relations cannot be assumed to be always in harmony and supports a greater degree of distinction between the two than Vogel allows. The problem here is that, as in many household work studies, Vogel's approach recognises some specificity for women's unpaid labour but adopts a framework which proposes close connections between unpaid and waged work to the point where the former is subsumed under the latter. The 'unified system' she provides is a unity built around one side of the equation, which flows one way: that is, the market drives the household. This is hardly an epistemological reconception of the limits of orthodox Marxism.[5]

By comparison, Young and Jaggar do not appear to offer a functionalist or unidirectional model. Their notion of a 'unified system' is constructed around a reconception of both public and private arenas in modern Western societies. Both postulate central overarching explanatory concepts intended to be applied to social totalities which presume that the capitalist market and the household cannot be analytically separated. While this type of unified approach does not dismiss the possibility of some differentiation between the market and household and can even allow for limited disjunctions between them (which is problematic in Vogel's writings), differences are de-emphasised by the use of overarching concepts. For example, in Young's use of the term 'division of labour' and Jaggar's use of 'alienation', the focus is upon connections between household and market to the point where they are conceived as part of one system and under the rubric of one central concept. Consequently the capacity to recognise different, even contradictory, trajectories for these domains is somewhat circumscribed in their work. Moreover, the 'unified' model is inclined to imply not only a singular analytic stance but a singular politics. The model is disposed to concentrate upon the interactions between sex and class hierarchies rather than their distinctive and therefore potentially disjunctive attributes.

That perception of centred unity or sameness can imperceptibly lead to a subsumption of the specificity of women's experience reminiscent of aspects of Vogel's work. The 'unified system' framework of Young and Jaggar, like that of Vogel, is somewhat less open to the polyvocalism of postmodernism and post-Marxism, or to any perspective acknowledging multiplicity and specificity, than frameworks which start from the point of differentiating distinct forms of power and social organisation. All-embracing categories that are deemed to be equally applicable to and offer explanations for every aspect of social life, such as the 'division of labour', contain the dangers of a singular viewpoint which has been the basis for marginalising women. Though Young explicitly intends to rework Marxism's singular focus upon class and its tendency to sex-blindness by providing a category (and theoretical model) which extends over class and sex, it is not certain that her version of singularity—which does not propose women as *the* subjects of the theory—can overcome either Marxism's hegemony or its masculinism. The sex neutrality of the overriding category 'division of labour' may therefore resurrect the phallocentrism and difficulties with 'sameness' of supposedly sex-neutral epistemologies like Marxism and liberalism. In common with these epistemologies, 'the division of labour' approach may be decidedly limited in terms of the particularities of women's experience and the specific logic of their labour.

## The continuing legacy of Marxism: unified system connections and constraints

Young's 'unified system' model is similar to the approach of many household work analyses and 'labour' studies in insisting that it is not possible to examine the household and the market separately. 'She attempts to critique what she sees as Marxism's acceptance of the public/private divide and to reintegrate these domains. Young (and Jaggar) promotes a unified analysis which rejects the pre-eminence of Marxism in feminist analysis and notes Marxism's masculinist evasion of the private sphere.[7] Nevertheless, in focusing upon the 'division of labour'—rather than on the Marxist account of class relations/waged labour—Young retains the Marxian inclination to privilege labour as *the* priority for social theory and feminism. A theory of labour/economics is still perceived in her approach as equivalent to a theory of social relations. Indeed, labour continues to be constituted as a causal category in regard to social relations per se and relations of domination in particular.[8] It is rather ironic in this setting that Young attacks Chodorow's analysis in part on the grounds that it is 'universalistic'.[9] Young criticises Chodorow's attention to the psychic because she believes that while the subject/psyche cannot be granted a universal significance, apparently labour can. Further, labour in Young's view is able to be historicised and is 'material'.[10] Whether or not the psyche is capable of historically specific interpretation, Young's concern with labour as *the* central feature of history/social relations and hence her identification of 'materialism' with labour and economic causality reveal more than a suggestion of an a priori priority given to Marxist accounts of base/superstructure and 'materialism' based in economic determinacy. I have indicated in Chapter 1 that these accounts are at least debatable for a feminist political economy but, beyond this, Young's acceptance of them scarcely supports her claim to transforming and overcoming the limits of Marxism.

Apart from the potential of such a neo-Marxist perspective to slide into a class-based model of sociality (which is in fact a difficulty in Young's work), the retention of the base/superstructure model and economic determinacy materialism does concretely result in the maintenance of Marxist and masculinist limits on the study of women's labour. This is particularly evident in Young's discussion of Chodorow, Harding and Hartsock and in her overall view of the place of 'gender identity' and psychic/psychological processes. Young argues that Chodorow perceives 'gender identity' (the development of masculinity and femininity) as *the* source of hierarchical sexual order and contends that Harding and Hartsock follow this approach. Young is concerned by what she sees as the loss or lack of a 'materialist' and sociohistorical perspective in this psychoanalytical framework. For her, the subject/the individual cannot be placed above the social and structural nor can the psychic be prioritised over labour. Chodorow's perspective is seen as flawed by individualism and idealism. However, even if one agreed that Chodorow sees 'gender identity' as the exclusive origin of male dominance—I do not consider that Chodorow's analysis is as clear-cut as this—the problem with Young's critique is that she remains bound to a Marxist view of base/superstructure and to its conception of materialism as equivalent to economic determinacy. Hence she can only recognise 'gender identity' as an individualist or 'ideational' question.[11]

Young consequently regards a whole series of women's labours in the private sphere, such as occur in husband-care and sexuality as well as in childcare, as somehow merely psychological and as superstructural effects of supposedly 'real' labours of other kinds. She fails to acknowledge, as Rubin and Ferguson do,[12] that sexual identity is a 'production' and involves a variety of activities that could quite easily be included under her narrow definition of labour. Moreover, Young relegates these activities to a secondary status and in doing so reduces the capacity of her framework to take account of the specificity of women's labour by implicitly tying the character and significance of labour to that which Marxism allows. Activities which epitomise the emotional economy, are most unlike 'productions' in the market economy, and highlight the particularities of sex relations, are deemed not-economic, not-labour and lesser in the analytical model Young proposes. The invisibility of women's labour (especially in the domestic arena) and of sex relations within Marxism is ameliorated in Young's 'unified system' theory but by no means overturned. In my view, Young's work may be criticised not simply because of her assumption of an exclusive and for-all-time economic determinacy but additionally because of the link between this assumption and elements of a masculinist vantage point which is blind to or de-emphasises significant aspects of women's experience and sex relations.

It might be possible to retain a version of the base/superstructure model which included both labours concerned with goods and services typically recognised by Marxism and household work studies, and labours associated with emotional/bodily/psychic–sexual identity 'production'. This possibility cannot be dismissed. However, a model that perceives labour as *the* priority in social analysis and economic determinacy exclusively as the most fundamental determination throughout history is subject to the critique of Marxism mounted in Chapter 1. There is a danger of slipping into a prioritising of class theory or of accepting dubious features of Marxism that arise in employing Marxist concepts like base/superstructure and economic determinacy (even when these concepts are employed within a feminist framework). Such dangers are evident in Young's work.

## The unified system critique of dual/multiple approaches

Young's critique of the 'dual systems' approach and her reason for promoting a 'unified system' model are based on a view that the division between the public and private domains in modern Western societies is problematic. Young focuses on the interwoven nature of class and sex systems and relatedly resists accepting a conception of separate spheres transposed upon the public–private divide. She conceives class and sexual hierarchies as part of one system and simply notes that the public–private split is a creation of capitalism. The separate spheres conception, which Young sees as central to dual systems theory, in her view involves taking as given a distinction drawn from bourgeois ideology.[13] One may doubt that dual systems theorists do indeed propose as complete a separation of sex and class, and relatedly the private and public spheres, as Young suggests, given that their approach may be seen as merely enunciating a two-phase methodology which first analyses the systems as distinct and *then* considers their intersection. Nevertheless, the dual system model is more inclined to emphasise separation than Young's approach, which could be a point for debate. But Young's rejection of 'separate' spheres rests upon a belief that the class system is the motor of a sociohistorical division that has been very important to modern Western sex relations—that is, the split between 'family' and 'economy'.[14] Her belief is quite in keeping with orthodox Marxist views of the dynamic constitutive power of capitalism and the comparative passivity and secondary status of the sexual order.

A historiography which attempts to discuss women's experience might be rather more sceptical about this privileging of class relations. Allen points out in this context that the conception of separate spheres and the association of domesticity with women pre-dates the rise of capitalism. She asserts that the argument that capitalism is the *cause* of the public–private division 'is both a faulty deduction and fundamentally unconvincing'. For Allen the domestic sphere cannot be regarded as a creation of bourgeois ideology since the organisation of similar sexually separate domains may be observed in a range of pre-capitalist Western societies.[15]

To say that capitalism *uses* domestic ideology discourses may be
descriptively and historically correct, so long as it is acknowledged
that its origin, cause and explanation is not the specifically capitalist
mode of production, but an historically longer struggle over power
between women and men.[16]

Young's inclination to underemphasise the public–private divide and
the distinctiveness of sex and class relations, when combined with a
framework which views the development of any degree of separation
as derived from one side of the equation alone, can only make one
doubt her attempt to transform the priorities of Marxism and its
limits in dealing with women's experience. Her concern to theorise
the *intersections* between class and sex, public and private, may well
be a useful antidote to the potential problems of drawing a firm line
between the public and private domains and the tendency to ignore
interconnections between them. However, Young slides into an
unwarranted dependence on the Marxist paradigm when she makes
interconnection *the* central focus and views that connection as driven
by the requirements of capitalism. Moreover, there is only a limited
sense in which Young's model allows for the household to 'drive'
the marketplace. The interconnection described is distinctly one-
sided: capitalism for the most part defines and determines the sexual
organisation of labour. While downgrading the historical agency of
sex relations, Young fails to appreciate the possibility of the incon-
trovertibility (or limited controvertibility) of power across the
public–private divide noted by Jacquette.[17] Young's unified and rather
one-sided model is unable to take account of research which suggests
that notions of connection between household and market cannot
be upheld as a matter of theoretical faith. England and Farkas, for
instance, emphasise that both the view of the market driving the
household and of the household driving the market may be challenged by

> reports of change in one sphere (women's increasing involvement in
> paid work) *not* being accompanied by anything like equivalent
> change in the other (involvement in household work). This odd
> phenomenon, as [England and Farkas] point out, amounts at best to
> weak connections between the two spheres and, regardless of the
> direction of the arrow one sees as the essence of the connection,
> must be regarded as a major puzzle for all theoretical positions.[18]

Such reports not only challenge 'unified system' models like that of
Young and orthodox Marxist approaches but also, as noted in
Chapter 3, household work studies which postulate a high degree of
connection between household and market and/or some notion of
'competition' between these spheres. The problem with all of these
accounts lies in their disposition to underplay the specificity of the

two domains which may inhibit the domains' capacity to connect or compete. Perhaps they are not same enough to be conceived in terms of functional linkage (however the direction of connecting flow is seen) or push–pull mechanisms.

Without doubt England and Farkas's point provides persuasive support for a dual or multisystem analysis, but on the other hand connection should not be dismissed. It is after all possible to construe the household economy as distinct yet allow for intersections with the market. As Thompson asserts from a Marxist perspective, social relations do not have to be viewed as inside or part of a particular economic system to affect that system.[19] Problems only arise if the effects or connections are inevitably taken to override, outweigh or marginalise the specificity of the economic system. Young, however, makes the mistake of presuming that connections can only be viewed through the conceptual lens of overweening unity/singularity. Thompson's comment demonstrates that it is possible to discern connection while at the same time proposing distinct systems. This may be useful to a feminist political economy if there is an accompanying concern to reject the privilege accorded the capitalist domain by the Marxist paradigm. Since this rejection is only partially achieved in Thompson and Young's work, their different accounts of connection are similarly constrained in the extent to which they can acknowledge the specificity of the household economy.

I am not suggesting that a feminist political economy should not deal with the intersections between 'household' and 'market' economies because of the likelihood of underemphasising the particularities of the former. I have noted, after all, the considerable advantages of stressing these intersections exemplified by household work studies which estimate the value of domestic labour in the market terms of money and time usage for the purpose of national income accounting and policy development. Furthermore, a perception of connecting links between 'household' and 'market' does not necessarily require repudiation of the distinct status of the household economy. Rather the question is whether connection is the major or sole focus, how the connection is conceived, and for what purposes. I do contend, nonetheless, that writers who give attention to intersections between household and market are almost invariably disposed to underemphasise the specificity of women's labour and sex relations, and commonly remain bound to aspects of phallocentric Marxism/liberalism. This is clearly an issue for 'unified system' models in which connection is the starting point, as can be seen in the case of Young and Jaggar's work.

## Unified system analysis and the equation of 'the economy' with the market

I have pointed out the ways in which Young's framework and overweening category 'the division of labour' continue in crucial ways to privilege Marxian concepts and in earlier discussions I have addressed similar problems in relation to Jaggar's central category, 'alienation'.[20] Both Young and Jaggar in their different ways give priority to labour and to human beings' relationship to labour. They retain a Marxian orientation to social determination and 'materialism' which influences their capacity to recognise psychic/subjective 'production' as labour, despite the fact that even socialist feminists no longer inevitably assert 'work'/'labour' (in the traditional sense) as *the* exclusive centre of feminist analysis.[21] Furthermore, the privilege accorded 'labour' is often associated with a residual tendency to construe capitalism as the only *economic* system. Though the particularities of women's labour may be acknowledged, the 'unified system' approach tends—in common with many other varieties of socialist feminism—to be uncertain about, underplay or ignore notions of an economic organisational form beyond that of capitalism. Hence Young still sees capitalism as the *system* of *economic* organisation and, not surprisingly, given her 'materialism', defines historical epochs in class terms as capitalist or pre-capitalist.[22] One is left with a sense that the lesser materialist status of sex in Young's approach may indeed marginalise its place in her social theory.

The difficulty of conceiving women's labour without any notion of a specific organisational form and the reliance on Marxian accounts of 'materialism'/economic organisation becomes more evident when Young describes capitalism as a type of patriarchy. Young attempts to evade her seeming inability to formulate labour in terms beyond those elaborated in Marxism by giving priority to class relations in her historiography but simultaneously declaring these relations to be indistinguishable from those of sex. If capitalism is a patriarchy, and the patriarchy which drives historical periodisation and the emergence of significant historical developments like the public–private split is capitalism, then it is hard to perceive any specificity for sex relations in this circular and empty effectivity. Reducing categories like patriarchy to capitalism or vice versa involves an all-embracing, totalising approach which is by no means limited to the 'unified system' model or to socialist feminism.[23] However it is more likely to occur in the 'unified system' framework and raises once again the difficulties of a kind of functionalism that cannot allow for the disjunctions or weak connections between the household and market economies suggested by Hartmann and by

England and Farkas. In this sense amongst others the 'unified system'
framework is probably inevitably limited in its capacity to acknowledge or analyse the specificity of women's labour and therefore has
implications for the employment of that framework in a feminist
political economy.

I have indicated that one aspect of the limits of the unified model
arises from its continuing debt to Marxist theories and priorities.
That debt is often evident in socialist feminist approaches—as I have
demonstrated through discussion of Young's writings—in the inclination to perceive only capitalism and the arena of waged labour in
terms of economic structure, organisation or system, and in the
associated tendency to be unable to envisage women's labour or sex
relations as having an economic system. Socialist feminists, along
with orthodox Marxist feminists, appear to have difficulty disentan-
gling themselves from Marxian assumptions and definitions
concerning what constitutes an economy or economics. For example,
despite a critical interest in women's labour, the editorial group of
*Feminist Review* and writers like Ann Curthoys and Anne Edwards
describe capitalism as the 'economic system', as equivalent to 'the
economy', and sex relations or patriarchy as somehow about
'power'.[24] Curthoys asserts that 'large aspects of gender relations *are*
economic, are bound up with the mode of production' and that
socialism is about communal ownership of 'the means of
production'.[25] 'Gender relations' are defined as economic to the
extent that they are connected to capitalism and the political aim is
to alter economic ownership of elements of capitalism. Even if 'means
of production' were interpreted to refer to technologies/modes of
labour organisation related to women's unpaid work, it is by no
means clear that 'ownership' of these 'forces' would be the critical
political task or would entirely transform expropriation of women's
labour.

A fundamental failure to conceptualise women's labour in anything other than the derivative terms of the market/class analysis is
evident in many socialist feminist accounts, reflecting an ongoing
dependence on Marxism in spite of an often simultaneous awareness
of the seriously deficient nature of the paradigm in dealing with
women's experience and activities. The 'materialism' of most forms
of socialist feminism amounts to a shift from *direct* usage of class
categories, concepts and priorities to describe sex relations (which is
a feature of orthodox Marxism and orthodox Marxist feminism) to
an *indirect* usage realised in the continuing employment of the
standpoint of capital in discussion of 'the economy' and the place
of women's labour within it. Thus, though socialist feminists frequently adopt more of a political economy perspective, focusing on
power in relation to both class and sex relations, they commonly

share the limitations of household work studies outlined earlier in the book. While women are usually closer to being viewed as the subjects—or at least one of the subjects—of analysis in socialist feminism than in many household work approaches, women's private labour is generally still not seen in its own terms. That labour is not conceived as the point of departure for a 'materialist' theory. Little attention is still paid to the question, What would a materialism look like which took the emotional economy as its vantage point?

## Dual system dilemmas: Barrett and Mitchell

The possibilities of such a question are only dimly realised in the 'unified system' model and, as I have noted, remain a marginal element of 'dual systems' approaches like those of Barrett and Mitchell, which postulate distinct analytical spaces for sex and class relations but do not recognise the sexual order as having a specific economic form or 'material' organisation. This version of the 'dual system' model shows that a conception of duality or unity does not by itself provide a basis for taking account of the specificity of women's labour and experience. Barrett, in common with many other feminist writers, points out that Marxism is an insufficient basis for the study of sex relations and that many aspects of women's situation are simply not reducible to the demands of surplus value. She therefore, from a socialist feminist perspective, dissents from those claiming a functional fit between sex and class and argues for a dual approach which construes sex relations as *analytically* distinct from those of class and proposes that there is nothing *intrinsic* to the latter requiring them to be linked to the former.[26] Such a stance might seem to place her at odds with Young's unified model, with its focus on the interrelationship between sex and class and a singular notion of struggle which refutes any perception that these social relations may require the development of different political weapons. However, Barrett's 'dual systems' approach is not as distant from Young's proposals as it first appears.

Barrett's dualism, like that of Mitchell, can be described as providing a 'non-materialist' account of sexual hierarchy combined with a 'materialist' account of class. She argues that sex relations are analytically distinct, yet historically embedded in a necessary reproduction of class relations. Terms like 'patriarchy' which allow for potential contradictions between these relations are prohibited as idealist and ahistorical because they cannot be specified for different 'modes of production'. Her view of the 'historically embedded' nature of sexual hierarchy returns her to a functionalist account of the relationship between sex and class which she has criticised. Indeed,

Barrett's critique of patriarchy rests on a singularly Marxist privileging of class categories. Patriarchy can only be seen as inevitably
'non-materialist' and ahistorical if there is a pre-given demand that
the periodisation of the relation of the sexes should synchronise with
or be defined in relation to 'modes of production'.[27] In other words,
Barrett's dual model is not so dissimilar from Young's unified framework in focusing upon the connection between the household and
the market at the expense of specificity, presuming the pre-eminence
of class in shaping history/historical epochs, and the inclination
towards a singular politics.[28] Barrett acknowledges the distinctiveness
of sex relations, but the analytical space she claims for this is
decidedly straitened given her perception that any account of those
relations that is not bound to Marxist procedures must be idealist
and cannot be reconciled with history per se. Barrett's implicit view
of Marxism as having a monopoly on 'historical materialism' must
delimit any question of a distinct sexual economy or a materialism
that places the particularities of women's labour at centre stage.[29]

Mitchell's 'dual systems' approach displays the same problems
with envisaging sex relations in materialist terms as Barrett's work
in that she explicitly conceives those relations in psychic/psycholog-
ical/ideological terms and, like Barrett, equates the 'economy' with
the class 'system'.[30] Mitchell's advocacy of differential logics for sex
and class allows for a greater awareness of the specificity of women's
position and for differential political strategies, but her analysis of
the domain of sex relations is curiously labour-less.[31] Thus her
approach, no more than Barrett and Young's almost exclusive emphasis on connections between sex and class, cannot easily produce a
feminist political economy which concentrates upon women qua
women and women's labour within a sexual order. Moreover, Mitchell in *Psychoanalysis and Feminism*, in common with Barrett, appears
to leave Marx's account of the class system relatively untouched.
Recognition of the distinctiveness of sex relations promised by their
versions of the dual system model does not appear to lead necessarily
to a fundamental reconsideration of economic categories and concepts. Young at least partially attempts this.

## Framing an agenda for a feminist political economy

My critique thus far suggests that whether or not feminist analyses
are inclined to offer a unified system or dual systems model, or
provide socialist or other perspectives, they do not in themselves
provide sufficient analytical choices upon which to frame an agenda
for a feminist materialism or a feminist political economy. What is
required is a stance which:

- does not prioritise points of interpenetration between sex and class relations ('household' and 'market' economies);
- allows for the possibility of interconnection but does not assume it (unity/connection and specificity are both conceded at theoretical and political levels);
- takes a methodological position for the purpose of the analysis which concentrates on labour yet does not prioritise it as *the* determinant of social relations;
- provides a feminist standpoint, conceiving women and sex relations as *the* subject of the analysis (consideration of specificity may well be viewed as the first phase of the methodology); and
- is sceptical of dependence upon frameworks which subsume women/sex relations under other classificatory categories (such as occurs in Marxism), thereby delimiting the investigation of the specificity of women's labour.

A feminist political economy/'materialism' does not require that all traces of theories like Marxism be expunged from its program. As I have noted, the standpoint of the market can indeed be potentially politically useful. Critics of socialist feminism like Elshtain have asserted that labours such as mothering cannot be regarded simply as the reproduction of commodified labour power.[32] Nevertheless, even radical feminists like Robyn Rowland, who is certainly not uncritical of Marxism, have used Marxian language to explore power and expropriation in sex relations. For instance, Rowland argues that male-controlled technology replacing women's reproductive capacity will leave women 'without a product' with which 'to bargain'.[33] Similarly, though MacKinnon characteristically draws attention to the distinctiveness of sex relations and women's labours and is inclined to reject 'unified system' models on the grounds that they merely incorporate feminism into Marxism, she is quite willing to place on record her continuing debt to the Marxist paradigm.[34]

The point is not that 'materialist' feminisms ought to be dismissed if they draw upon Marx, or that use of Marx or any other theorist who does not focus on women/sex relations is inadmissible, but rather to be wary of perceiving women/sex relations in ways that cannot do justice to aspects of the specificity of women's experience and the relation of the sexes. In MacKinnon's view there are dangers in a dependence on Marxist content, method and priorities in arenas in which Marx did not have insights.[35] This corresponds to my own argument that when one looks at women's private labour, and the economic organisation of that labour, a Marxist standpoint involves the loss of some features particular to these arenas. Since such features may be crucial to a feminist account of political economy, Marxian economic categories, concepts and sites of analytical privi-

lege must be reconsidered. This means avoiding incorporation of
feminism into Marxism by questioning approaches which prioritise
labour and thus centrally perceive women as workers, which is a
difficulty in 'unified system' models.[36] It also suggests a critique of
analyses which are unable to construe any distinctive materiality/eco-
nomic organisation for sex relations which occurs in both 'unified
system' and one version of 'dual systems' models. Furthermore,
approaches which continue to regard capitalism/class relations as 'the
economy', and those which tend to assume that Marx's conception
of labour—even within analysis of capitalism—may be left intact,
should be treated with caution. The point is not whether Marxism
ought in all respects be abandoned (or at least not necessarily) but
that there is a need for a political economy that is not *derivative* of
Daddy's. A political economy is required which challenges, as
O'Brien suggests, the fundamentals of theories that not only subsume
women within sex-blind paradigms but are phallocentric[37], especially
in their inability to conceive labour from the perspective of the
private domain and women within this.

The starting point for such a political economy is therefore to
move beyond the interpretation of women and sex relations through
sex-blind/phallocentric economic concepts and categories—that is
sexual economyths—to propose, in Jaggar's terms, women's unique
position as the basis of theory[38], and to develop a theory from the
vantage point of the domestic/private realm in modern Western
societies.[39] This feminist political economy does not ignore 'unified'
forms of analysis that focus on the relation of the 'household' and
'market' economies. It merely asserts that this relation *can* be examined from a perspective that *begins* with the notion of the specificity
of women in the household, that is, both unity/sameness and diver-
sity/difference can be viewed from a methodological centre stage
located in women/sex relations and the private sphere. Without the
development of such a standpoint, economics remains bound to a
male and public analytical framework by default and thus cannot
adequately deal with certain elements of modern Western labour.

The feminist political economy outlined is no more an entirely
totalising picture of economic processes than is the economics
founded in male and public activities. Meredith Edwards points out
that different economic 'measures' are appropriate for different
purposes.[40] Similarly, different analytical stances are relevant for
different theoretical and political intentions. If feminism has a concern to understand and explore sexual forms of power and
expropriation, one important method for doing this is to conceive a
woman-centred study of labour and economic forms, to examine the
idiosyncrasies of a *sexual epistemology of economics*.[41] To take
women/sex relations and the private household as the starting point

for a theory of political economy precisely confronts 'male-stream' assumptions in economics, whether these are of a 'mainstream' (neoclassical/social liberal/social democratic) or Marxist variety, and highlights power relations between the sexes. The particular adequacy of this political economy to the aforementioned purposes provides a compelling reason for its development. It is no coincidence that when Hartsock begins to outline 'a theory of the extraction and appropriation of women's activity and women themselves', her discussion of what is distinctive about the organisation of labour and power linked to sex relations almost exclusively draws upon examples from the private sphere.[42] The description of sexual power relations and expropriation within them rests heavily upon the specific character of the 'household economy' as against that of commodified labour. I suggest therefore that to deal adequately with women/sex relations in a feminist political economy it is in fact *necessary* to *centre* upon the private domain. It is not just a question of an alternative perspective among many.

Yet it is striking that not only has there been relatively little research in the field of feminist theories of power in private labour relations compared with those covering waged labour, and remarkably few studies of the household economy/household work[43], but almost no writings dealing with women, sex relations *and* household labour have proceeded from the point of view of the politics of the private household. Though it is almost unimaginable that examination of men's labour in modern Western societies would begin by looking at their unpaid labour, such a peculiarly reversed methodology is commonly adopted in relation to women's labour; that is, women's labour is invariably perceived in relation to waged labour and through the lens of the public domain despite the massive contribution of women to private work, the considerable contribution of that work to the total sum of economic activities undertaken in national economies, and women's lesser contribution than that of men to waged labour. Obviously the concentration of research on women's labour from the vantage point of the public sphere and the market has been driven by the hegemony of 'exchange' models in economics that Hartsock has noted[44], the associated dominance of neoclassical/Liberal and Marxist forms of analysis[45], and the masculinism of both 'exchange' frameworks and social theory. Beyond this, Millman and Kanter have distinguished a number of themes in the social sciences generally which can be said to demonstrate that the vision of social life in research studies has been limited by a failure to take account of, let alone view as a priority, the private domestic sphere and the experience of women.[46] These influences will tend to mean that even feminist accounts of economics and women's labour are likely to adopt a public 'market' oriented approach. They too will be inclined not to develop an economics from the viewpoint of the private domain, in spite of their criticisms of economic theory and the likelihood that such a viewpoint is essential for explorations of sexual power and sexual labour relations.

Studies such as those by Mumford, Sharp and Broomhill, and Waring demonstrate the disposition to remain largely within the confines of conventional economic paradigms. Though her book is titled *Women Working: Economics and Reality*, Mumford's work is entirely restricted to analysis of 'the labour market' and 'labour market' theories, with no discussion of private economic organisation.[47] Both Sharp and Broomhill's and Waring's works attack mainstream economics and insist that women's unpaid economic contribution be recognised, but the recognition they outline is in the form of *including* women's labour in a framework dominated by the perspective of the public arena. Domestic labour remains a marginal and 'deviant'/particular field. Hence Sharp and Broomhill conceive a 'feminist economic strategy' that does not mention even the possibility of an economic paradigm which proceeds from the specificity of the sexual labour 'system' in the private domain. Rather their strategy centres upon the connection between the household and market economies. There is an implicit tendency to identify 'the economy' with capitalism and to discuss the sexual division of labour in terms of a dimension of capitalist society.[48] Relatedly, Waring's account of unpaid labour asserts the need to include estimates of its value in economic measures but at no point acknowledges any limitations in the derivative 'market' indices she outlines for this purpose.[49] The inclusive approach adopted by these writers is directed towards the important political task of making women count, making their labour visible within existing economic theory. This is without doubt a particular purpose which offers considerable advantages to women. Nevertheless I think that this visibility is limited and bought at a cost—as it is in the case of household studies: the cost of continuing to de-emphasise certain of women's private activities, especially those related to emotional labour, and the specific form of sex relations. There remains a place for what is not heeded in these analyses, namely a visibility based upon considering women's private labour in its own terms.

The methodological position of focusing on women/sex relations and the household/domestic sphere in modern Western societies is encouraged and even demanded by the overriding reach of perspectives which marginalise or ignore any stance that does not perceive the public domain and—implicitly, explicitly or by default—men as the subjects of analysis. Harding asserts in this setting that '[e]xisting bodies of belief do not just ignore women and gender; they distort our understanding of all social life by ignoring the ways women and gender shape social life and by advancing false claims about both women and gender'. Consequently, she says, researchers cannot simply add women and sex relations to existing epistemologies.[50] The project of rehabilitating these epistemologies, which inform household work studies, 'unified system' accounts, 'dual systems' approaches so far discussed and feminist works like that of Mumford, is undoubtedly useful but cannot evade the limits of a masculinist/market analytical location. Furthermore, the claims of a women/sex relations/household perspective are supported by Phillips's point that unless there is some deliberate attempt to privilege women, the hegemony of masculinist thought may easily be resurrected: '[u]nless we discuss explicitly what any strategy means for women, we fall unthinkingly into policies for men'.[51]

This is not, of course, to promote the notion of an epistemology untainted by phallocentric impingements[52] or without recourse to theories like Marxism, since the first suggestion involves an impossibility and the second refuses the possibility of particular political contexts in which use of masculinist and 'market' oriented themes might be advantageous. What is allowed by focusing upon the distinctive position of women in the private sphere is, as noted earlier, a mechanism for confronting in a more thoroughgoing sense the frameworks and presumptions of such theories, as well as providing the opportunity for a way of construing social life that has as its primary political purpose discussion of sexual power relations. The political benefits of such a procedure cannot be underestimated. For instance, the standpoint of women/sex relations/household can offer a political strategy whereby women's private activities might be seen as a basis for measuring or reconceiving the values and organisational forms of the public/market domain,[53] rather than accepting that political/theoretical aims always take the path of assessing and reconceptualising the private sphere in the terms of and for 'inclusion' in the reference point of the public/market domain. The claims of feminism as a social theory and social movement are enhanced by a program whose axis is found in women's experience.

Additionally I would note that, even according to crude singular measures like time usage, the expropriation of women's labour *cannot* be adequately investigated without a focus on their private work. Walker's 1967–68 Syracuse study—which involved a comparison with research work from the 1920s, and Sacks's examination of several Soviet Union cities for 1923 and 1966, both showed increases over time in women's work time per week far above those in men's work time, despite the fact that 90 per cent of women in the Soviet Union (in 1970) were participating in the waged labour force.[54] The Marxist political aim of bringing women into waged labour and the Liberal program of achieving 'equality' in the public domain clearly have some problems insofar as women's participation in waged labour intensifies their oppression and relatedly fails to equalise men's and women's labour in the household. But more than this, studies like those of Walker and Sacks substantiate the point that the connections between the household and market economies are weak: women's contribution to household work is little diminished or altered by market labour. Hence epistemologies which merely include women in a primarily market-oriented approach, and hence fail to acknowledge the distinct economy and specific logic of the private sphere, must also fail to grasp the extent and character of sexual expropriation. A critical feature of feminist political economy must therefore be recognition of dual or multisystems, of more than one 'economy', and the particularity of the 'household' economy. While I have pointed out that adoption of a dual/multisystem approach alone is not sufficient for a feminist economics, when this is combined with a recognition of a distinct sexual economy 'a revolution in epistemology' becomes a definite possibility.

It is indeed very difficult to find a justification for the continuing almost exclusive reliance on paradigms which try to explore and explain intensification of expropriation of women's labour over time and massive differences in men and women's work time—Sacks found that the women in his study worked 17 hours a week longer than men[55]—by indirect, derivative methods that are based 'outside' of the site where their oppression resides and refer to a logic and labour forms which are typically acknowledged not to be the same as those found in the household. A feminist political economy may draw upon market paradigms and indirect methods, but it cannot afford to depend on these. There can be no excuse for not attempting a *direct* analysis of sexual economies and thus a political economy which begins with women, sex relations and women's private labour. I have given an account of some of the problems faced by writers dealing with women's labour and feminist 'materialism' and suggested aspects of an agenda for a feminist political economy.[56] Now it is appropriate to turn attention to those feminist theorists who have proposed models that start from the specificity of the 'domestic economy' and discuss their attempts, and those of others, to clarify the nature of expropriation in this economy in the light of the agenda so far outlined.

# 5 Dual/multiple vision: clarifying the perspective of a feminist economics

Some respite to husbands the weather may send,
But housewives' affairs have never an end.
(Thomas Tusser, Preface to the *Book of Housewifery*[1])

'Look at this', cried Father, flicking the paper with the edge of his finger.
'Germ warfare, atom bomb, hydrogen bomb. That's *all* you read!'
'Personally,' said Mother, 'I've a big washing this week.'
(Ray Bradbury, *The Last Circus*[2])

I have proposed the development of a *sexual* epistemology of economics which differs from the masculinist/'market' preoccupations of most forms of economic theory and analysis, including many feminist approaches, and which is intended to explore criteria towards the possible elaboration of a 'materialist' feminism. The critical elements of this agenda outlined in previous chapters involve a dual/multisystem framework dealing with modern Western societies and postulate a distinct economy/'materiality' residing in the household/private domain. The effect of such a conception is not to assume connections between the household and the market (that is, a notion of *unity*), but to perceive connections from the standpoint of the household, and to pursue these connections only in the second stage of the methodology—the initial focus being upon the specificity of the household. An associated aspect of this focus is the crucial concern with women and power in sex relations. The economic theory espoused is a sexual political economy, a sexual politics of women's labour.

In the above statements there is a manifestly cautious attitude to epistemologies like Marxism which do not operate from the standpoint of women, the relation of the sexes, and the household, and which tend to identify the market/public domain/waged labour with 'the economy'. Moreover, the agenda proposes a critical view of assumptions drawn from these epistemologies, such as the notion of the priority of labour/economics in social theory, and of approaches which are disposed to accept that such epistemologies are adequate, even in their own terms. Women's work, epitomised in this account

75

by women's *private labour*, is conceived as *differing systematically* from that undertaken by men and from the imperatives of commodified labour. The degree to which even women's waged work is distinguishable from men's labour is taken for the purposes of this political economy as likely to be derived largely from the specific logic of the 'household' or emotional economy. In other words, the commonly accepted vantage point of economic theory is reversed and the hegemonic authority of 'mainstream' (neoclassical/social liberal/social democratic) and Marxist approaches is challenged. This is not to suggest that the household drives the market, because what is crucial here is the notion of distinct power systems. However, such a perspective on connections between the two domains at least in relation to the arena of the *sexual organisation* of labour (which exists in both domains) is probably encouraged within the framework described.[3]

### Possible problems: political/theoretical marginality and the question of commonality among women

To this point I have concentrated upon the advantages of a standpoint which prioritises women, sex relations, and the household/emotional economy in the private sphere. The methodology initially at least centres on the *specificity* of women's private labour rather than its similarity or connection with other forms of labour (unity). There are some difficulties and political dangers associated with this approach. It is historically contextualised in the sense of a central concern with modern Western societies and is open to further clarification for particular settings. Therefore it does not imply that the specificity of women's private work is in any sense inevitable or that this specificity originates from or reflects some essential romanticised characteristics of 'Woman'. Nevertheless the focus on the distinctiveness of modern Western women's domestic labour (as against that undertaken by men and as against the imperatives of waged labour), precisely because it involves a critique of market perspectives, may suggest a glorification of separation. As Cocks has argued in relation to radical feminism, the 'household'/private sphere standpoint could appear to doom the study of women and sex relations to marginality, to a spatial location and emotional logic that perpetuates women's otherness. In Cocks's view, the celebration of specificity can be enslaving, not emancipatory. Attention to separation/distinctiveness may serve to replicate a state of marginal rebellion which leaves men's pre-eminence intact and thus serves to cement male dominance rather than a feminist political program. She states that 'the dominant culture' and the 'countercultural' perspective of radical feminism, in attending to women's otherness, 'engage in a curious collusion, in which . . . a rebellious feminism takes up its assigned position at a negative pole'.[4]

This is certainly a powerful rejoinder to my own previous arguments concerning the political dangers of frameworks which do not take sufficient account of specificity. Cocks provides persuasive grounds for welcoming accounts like those of Waring, Sharp and Broomhill, and household work studies, whose method for making visible women's private labour and sex distinction is to *include* these elements in a primarily market-oriented economic theory and market indices of 'value'. On the other hand, while Cocks shows the dangers of an *exclusive* focus on specificity and the *necessity* for any approach employing this focus not to promote such a degree of separation that theories of the market/public domain remain untouched, she assumes that emphasis on specificity inevitably leads to entrapment at the 'negative pole' of theory and politics. Though the pre-existing marginality of women's experience and the private domain does mean that approaches drawing upon these fields will be conceived by 'the dominant culture' as of little significance, if no perspective is ever advanced which precisely claims otherness and transforms it into the centre—the subject—of analysis, then the 'negative pole' will always be negative. Otherness may well be transformed by *inclusion* in the conceptual scope of 'the dominant culture' and I am certainly not suggesting that this methodological program be abandoned. However, I remain unpersuaded that this inclusion can challenge phallocentric assumptions any more than can an analytical location around notions of specificity. Indeed, it seems to me that the programme of inclusion could be less successful in this challenge than the latter approach. Marginality continues in inclusive approaches insofar as the basis for inclusion generally lies in the acceptance of a 'deviant' status for women's experience and of a delimited recognition of features of women's experience that do not correspond to dominant paradigms. I remain uncertain how areas of women's private work such as 'emotional labour' can be adequately examined without focusing on the distinctiveness/separation of that work, since critical elements of emotional labour simply *cannot* be registered by market indices.

But even if a women/sex relations/household standpoint does not necessarily entail glorification of 'Woman' as other or political separatism to the extent of ineffectiveness, the perspective of the particularities of women's labour does raise further analytical/political questions. The approach contains both the advantages and problems of a concern with what is common to women in modern Western societies. While its strategy places commonality (unity/sameness) between the household and the market as a second phase of the methodology and focuses upon diversity/specificity initially, the approach has an overarching conceptual framework which registers commonality among women. Diversity is not inevitably disregarded but the centrality of the category, 'women', may suggest the danger of rendering invisible issues of class and race, for example, and raises the critique of homogeneous categories of the subject proposed by postmodernism in particular. While the postmodernist attack upon the notion of the subject and located subjects, including sexual identity, may be seen as problematic on a number of grounds,[5] to the degree that this critique converges with doubts about a 'unitary' feminist approach—exemplified by the scepticism of 'women of colour', there does appear to be a basis for caution. bell hooks in this context emphatically states in *Feminist Theory: From Margin to Center* that feminism is not founded in the shared experiences of women, for women's experience of male dominance differs according to race and class among other social variables. Feminism, in hooks's terms, draws instead upon women's common *resistance* to these differing forms of male domination.[6]

In my view it is possible to concede much of this critique without abandoning an analytical stance proposing some notion of commonality among women, which can be placed alongside hooks's framework of political solidarity. hooks's depiction of the assumption of shared experiences as associated with white feminists' failure to acknowledge the specific history and experience of black women is of considerable significance in an analysis which focuses on a conception of specificity. Nevertheless I believe her rejection of commonality in *Feminist Theory* (published 1983) may be overstated. Alternatively, it may be argued that in more recent works hooks intends her rejection to apply to some but not all aspects/accounts of commonality.[7] Whatever the interpretation of her viewpoint, there is a useful strategic point in hooks's strong doubts regarding commonality related to the comparative marginality of race as a framework for theory and politics and to the hegemony of white feminism. In this context, while I recognise the need to challenge completed and uncritical assumptions regarding women's commonality, analysis of women's private labour in modern Western societies—in household work studies for example—does seem to me to reveal *certain* commonalities experienced by women across different classes and races.[8] I suggest that it is at least worth considering the implications of this; though at the same time taking from hooks's work the critical argument regarding diversity amongst women and stressing that commonalities must be examined in the light of particular contexts. Differences between women's experiences related to class and race positioning do not have to obliterate any conception of consistency and can be upheld simultaneously, in my opinion. It is not a question of dichotomous perspectives in which the choices are only either/or. As Nancy Cott has observed regarding the complexity of feminism, '[feminism] acknowledges diversity among women while positing that women recognize their unity. It requires gender consciousness for its basis, yet calls for the elimination of prescribed gender roles.[9]

Secondly, this approach parallels a point discussed at the end of Chapter 2. In Harding's account of the problem of unity versus diversity amongst women she outlined two solutions, one of which was the development of a feminist standpoint epistemology and the other, political solidarity. Harding appears to see these as alternatives, but I have argued that they might be combined in some ways.[10] It might be possible to undertake an analysis which proposes a *feminist standpoint*, focusing upon commonalities between women and recognising differences within these commonalities, yet which at the same time advocates a notion of *political solidarity* based in a federation of feminisms resisting all the ways in which male domination is enacted, both common and different. Though I can easily see the problems of perceiving social life through the lens of a unitary or homogeneous conception of women and sex relations, homogeneity/totality can be construed in a variety of guises and as having various 'scopes'. Its use is not without political consequences or dangers, but then the attack upon *any* employment of homogeneity is not politically innocent either.[11] Some judgment must be made as to when the constitution of commonalities may prove advantageous. A critically informed account which places women and sex relations at centre stage appears to me to be not only theoretically/empirically justifiable and even necessary, but politically strategic. Harding's insistence that this account should be considered as crucial to a feminist epistemology at this historical juncture is relevant here.

> Should women—no matter what their race, class, or culture—find it reasonable to give up the desire to know and understand the world from the standpoint of their experiences *for the first time*? As several feminist literary critics have suggested, perhaps only those who have had access to the benefits of the Enlightenment can 'give up' those benefits . . . [I]t is premature for women to be willing to give up what they have never had.[12] [emphasis in original]

One suspects that bell hooks would not entirely disagree. Her criticism of commonalities and the category 'women' provides an important qualification to the adoption of a feminist standpoint but is not, I think, fundamentally antagonistic to it. The doubts regarding commonality expressed by many 'women of colour', women from working-class backgrounds, and feminist postmodernists regarding commonality are perhaps most usefully employed in rejecting an exclusionist commonality, a narrow and distorted homogeneity, which forecloses diversity *within* commonality, and *alongside* it in the arena of political solidarity.

## Further clarifications regarding dual/multiple system frameworks

Having outlined aspects of the definition of sexual epistemology of economics and indicated potential difficulties in the approach, I now turn to those feminist 'materialists' who have undertaken analysis from the point of view of women, sex relations and the distinctive character of the 'household' economy and who might seem, at least initially, to proffer frameworks suited to the agenda so far developed. These feminist writers propose a specific economic form/'materiality' residing in the household/private domain. They encourage the fullest acknowledgment of the particularities of women's private labour by asserting that this distinct materiality should first be analysed separately (and therefore presumably in its own terms) before connections between it and other economic organisational forms can be discussed. Additionally, such approaches are sceptical of epistemologies like Marxism which do not operate from a feminist standpoint and the vantage point of the household/private sphere. These feminists espouse a version of dual systems (multisystems) theory in which the social relations of both class *and* sex (at minimum) are deemed to possess (distinct) material/economic structures. Relatedly, the two forms of social relations are not considered to be 'two heads of the same beast. They are different beasts, each of which must be fought with different weapons'.[13] In other words, this version of the dual/multiple systems perspective postulates that accounts of a singular economic system or one struggle are erroneous: a dual (or more probably multiple) materiality and politics is advocated.

Feminist writers such as Eisenstein, Chodorow, Harding, Hartmann, Hartsock, Delphy and Ferguson might be included under the rubric of a dual/multiple materialist approach. However, given the criteria I have suggested as useful for a feminist political economy, it would seem that the work of these feminists is subject to some of the same problems discussed in relation to household work studies, 'unified system', and non-materialist/ideological sex relations versions of 'dual systems' theory. Despite suspicion of the hegemonic analytical authority of Marxism, the tendency to see Marxism as Master-theory and feminism as the junior partner[14] is frequently revisited in dual/multiple materialist feminisms. For example, the pervasive determination to conceive sex relations in terms of a causal and functional connection with those of class is evident even in accounts which rest upon a view that the relation of the sexes constitutes a distinct system analytically, materially and historically. While dual/multiple materialist feminisms should logically represent a perspective which allows for conflict/disjunctions between class and sexual systems, some writers such as Eisenstein and Chodorow appear to be more closely linked to a class reproduction model than might be expected. The one-sided materiality of writers like Barrett is also resurrected.

Eisenstein states: 'I choose the phrase capitalist patriarchy to emphasize the mutually reinforcing dialectical relationship between capitalist class structure and hierarchical sexual structuring. Capitalism uses patriarchy and patriarchy is defined by the needs of capital.'[15] Class and sex here appear indistinguishable rather than distinct: connection is exclusively privileged over specificity. Moreover, Eisenstein's use of language is revealing. Capitalism uses patriarchy, while patriarchy is *defined* by capitalism. Both phrases depicting the 'dialectical relationship' characterise sex relations as passive. The disposition towards functionalism and a potential or overt prioritising of Marxist theory is also evident in Chodorow's work. She argues that there is nothing *intrinsic* to capitalism that requires it to be linked to sexual hierarchy. Nevertheless, while noting the lack of any necessary or logical connection, she asserts that the sexual division of labour 'has certainly been convenient to capitalists'.[16] The assertion of convenience may not be so very distant in practice from Eisenstein's account of a supposedly mutually reinforcing relationship, and tends to avoid the question of whether sex relations have actually been all that well suited to capitalism. Though it may be that these systems have conveniently or otherwise reinforced one another, the option of disjunctions/conflict is not present in these analyses. It is difficult to see how a conception of distinctiveness or specificity can be maintained in the face of complete harmony. At minimum, the straightforward collusion of 'interests' proposed undermines any theoretical/political significance to the material particularities of the sexual system.

This point is underlined when one notes that Chodorow's account of sex relations is inclined to depend heavily on a psychic/psychological framework which is only weakly and rather vaguely linked to a distinctly sex-based material/economic organisational form. I do not agree with Young that Chodorow sees psychology as *the* cause of male dominance.[17] Chodorow is not in any simple sense an 'idealist' since she does describe *some* relationship between a sexual division of labour, essentially defined in terms of childcare and psychological aspects of sex relations. Furthermore, the 'production' of sexual identity itself involves labour, as I have noted elsewhere. The relationship between the economic structure of sex relations and psychic/psychological processes in Chodorow's work is unclear and perhaps circular, insofar as she implies that the latter leads to the former and, less obviously, vice versa. When she argues that social, material changes in the sexual division of labour can break the psychological pattern of sexual identities which promotes male dominance (she particularly refers to shared parenting in this context), she does not depict a link between psychology and economics which is exclusively in one direction and delivers no manifest final aetiology. Nonetheless, Chodorow does rely upon a dehistoricised psychoanalytical framework. The analysis of economics in the sexual system is sparsely developed and marginalised, despite her political program for social change in the area of women's labour.[18] The association between the transhistorical psychoanalytical model and presumably historically specific sexual divisions of labour is explained only to a limited degree. Her discussion of labour appears to take on a psychological complexion in that the psychoanalytical focus overshadows other elements of the approach.

I would suggest that her overriding emphasis on psychic/psychological mechanisms makes Chodorow's somewhat functionalist appraisal of the relationship between class and sexual systems all the more worrying and that this functionalism is not a coincidence given her restricted capacity to explore a distinct materiality in the sexual system. The failure to conceive fully an economic structure for sex relations is translated imperceptibly into an association of economics and labour primarily with class relations. In addition, to the degree that dual/multisystems theories tend to privilege connections at the expense of disjunctions between class and sex relations, they are also inclined in practice to marginalise and circumscribe any account of materiality/economics in the sexual system. They in fact tend to replicate the problems of dual systems approaches which perceive sex relations as entirely or primarily ideological/psychological/non-material, identify capitalism with economic processes, and by default leave the sway of Marxism in the field of labour comparatively unchallenged. Not surprisingly, Chodorow's analysis collapses towards that of Mitchell.

The disposition of dual/multiple materialist models to slide away from a developed and dynamic account of an economic form for sex relations, and in the direction of frameworks which are unable to disentangle Marxism/class relations from materialism, certainly indicates the enormous difficulties involved in conceiving economics from a position outside of dominant epistemologies in the field. It shows also the complexities of formulating a 'revolution in epistemology', which Harding proposes in order to evade an over-reliance on inadequate existing modes of thought.[19] This problem is evident in Eisenstein's work and paradoxically in the writings of Harding herself in relation to economics. Though Eisenstein advocates a dual model

which appears to grant sex relations a specific materiality and
effectivity, her discussion of the 'dialectical relationship' between sex
and class must be set alongside her statements that patriarchy is not
reducible to 'the economic system'.[20] Given that Eisenstein is proposing a materialist analysis, it would seem that the dynamic of sex
relations is markedly undercut by a perspective which largely constitutes those relations as a relatively autonomous ideological aspect
of a class-based economics. The approach remains derivative of
Marxist conceptions of social life and follows Chodorow's inclination
to slide towards dual system models more like those of Mitchell and
Barrett.

Harding, in advocating a feminist standpoint based in the specificity of women's experience and sex relations, would also appear
to support a dual/multiple materialism, yet she too stumbles when
it comes to reconceiving epistemologies in the arena of economics.
Like Eisenstein, Harding does not quite disassociate herself from
acceptance of the equation of Marxism/class theory/capitalism with
economics per se. She states, without noticing the irony, that a
feminist standpoint which conceives nature and social relations from
the perspective of women's situation,

> sees sexual relations as at least as causal as economic relations in
> creating forms of social life and belief . . . In contrast to Marxist
> assumptions, [women] are not merely or perhaps even primarily
> members of economic classes, though class . . . also mediates
> women's opportunities.[21]

In her attempt to refute the hegemonic claims of Marxism and to
espouse the particularity of sex, Harding reasserts the non-economic
status of this classificatory category. This must severely delimit the
program of a feminist standpoint for any materialist theory as well
as curiously returning to Marxism its dominance over economic
epistemologies.

## More specific perspectives in the dual/multiple system field

Dual/multiple materiality is consistently elaborated by writers like
Hartsock, Hartmann, Ferguson and Delphy, and thus they, more than
any of the previous household work and feminist approaches previously discussed.[22] are able to give attention to the character and
extent of specificity. I have argued in earlier chapters the necessity
and usefulness of exploring the particularities of women's labour and
of not restricting the analysis of this specificity by derivative, inappropriate and pacified/marginalised frameworks. Given this, the four
feminist writers mentioned may provide a beginning for a sexual

epistemology of economics. Hartsock, Hartmann, Ferguson and Delphy propose a dual/multiple materiality in which sex relations are regarded as having a distinct economy.[23] This may be explained by reference to Rubin's coining of the term 'sex/gender system' which refers to a 'neutral' classification like the Marxian 'mode of production'. The term locates a system which in all societies deals with sexuality, sexual subjectivities and the rearing of children, and consists of a set of relations 'by which society transforms biological sexuality into the products of human activity'.[24] Moreover the 'neutral' term 'sex/gender system' allows a view of oppression in sex relations as a particular set of arrangements (patriarchy) within the domain previously described. This system, whether patriarchal or not, is distinct from the Marxian 'mode of production' in that it is neither reducible to it nor significantly inevitably in harmony with it. Like the 'mode of production', the 'sex/gender system' is viewed as having a distinct materiality, a political economy, which is found in the sexual division of labour. In patriarchal societies the sexual division of labour takes an oppressive form.

Hartsock, Hartmann, Ferguson and Delphy all draw upon this account of an historically specific patriarchal sexual division of labour in some way and relatedly regard men's control over women's labour as central to a feminist political economy. To a greater or lesser degree they elaborate the notion of a distinctive economic system/specific organisation of labour within the arena of sex relations, though they may differ over whether the private domain is a site, the critical site, or the originating source for a sexual economy. Delphy tends to describe this organisation in terms of men's control over women's labour in the institution of marriage, at least within modern Western societies. She regards marriage as framing a distinct 'production' mode coexisting with capitalism. This provides a conception of materiality for patriarchy/sex relations in which marriage is conceived as an economic unit headed by a paterfamilias and in which the husband appropriates the labour of his wife in the domestic sphere.[25]

The approach implies considerable similarities between capitalist and precapitalist societies with regard to a specifically patriarchal mode of organisation of labour. There is no pre-given requirement that its periodisation synchronise with Marxist accounts of class epochs.[26] Delphy also refers, rather inconclusively, to indirect, more generalised mechanisms of appropriation in the institution of motherhood. The link between the organisation of labour in the institution of marriage and motherhood is somewhat unclear but Delphy clearly indicates a specific logic for that labour, a distinguishable economy that is defined by its relational/emotive character.[27] Furthermore, her approach offers an alternative to analyses based in the 'household' insofar as she promotes a politicised reading of labour relations founded in power and expropriation. Delphy's political economy cannot be confused with discussion of a 'unit' or 'sector': the economy she discusses, though it is located in the private domain of modern Western societies, is not centrally defined by a domain or site. Its character is predicated on men's control over women's labour, which happens, in her view, to *originate* in the domestic sphere. Delphy argues that women's position in wage labour is the consequence and not the cause of their position in the institution of marriage.[28] Even if this account of a historically specific critical institutional site in the private domain is debatable, Delphy undoubtedly offers the possibility of a sexual epistemology of economics from the standpoint of women, sex relations, and the 'household'/private sphere.

Connell has criticised her approach precisely on these grounds. On the one hand he acknowledges that her analysis of marriage (and motherhood) enables a conception of both a sexual organisation of labour and a dynamic for a sexual logic of 'accumulation' in its own right. However, he argues that Delphy's confinement of this discussion to 'the household and the one-to-one marriage relation' misses the 'larger possibilities of accumulation in industry'. In his view, 'the scale of economic inequalities resulting from labour organised through marriage is severely limited, when compared with the economic inequalities that can be produced by accumulation in industry'.[29] The first problem with this critique is that Connell slips from a discussion of sexual economics that appears to take into account the meaning of inequalities in input/output and in organisational power and control over labour in the broadest sense, to a much more limited meaning revolving around economics as equivalent to monetary measures. His use of the term 'accumulation' is significant here. Secondly, Connell ignores, or rather does not accept, Delphy's insistence that the sexual organisation of labour and expropriation (rather than 'accumulation' with its Marxist/class overtones) is not of the same order as class relations and the mechanisms of monetary/capitalist accumulation. In Delphy's approach sexual economics cannot be seen as a dimension of capitalist accumulation, as Connell appears to propose. Indeed Connell makes use of a familiar circular analysis which tends to reduce the specificity and effectivity of sex relations to a feature of a 'larger' class-based framework and focuses upon harmonious interconnections between sex and class, while supposedly accepting some measure of historical independence for sex relations. He states in this context that '[g]ender divisions are a fundamental and essential feature of the capitalist system . . . Capitalism was partly constituted out of the opportunities for power and profit created by gender relations. It continues to be.'[30] Once again, capitalism and patriarchy are virtually indistin-guishable. Allen, reflecting Barrett's attack on this kind of position, notes the contradictory and rather empty effectivity this implies for sex relations. The position postulates that capitalism is a patriarchy and that the patriarchy to which analysis should attend is capitalism, and yet, at the same time, patriarchy is posed as somehow external to capital.[31]

Connell's assumption of the relatively small scale and narrow vision of the household/marriage standpoint reflects a commitment to Marxism and the standpoint of the market as much as anything else. He does not appear to be aware that just in crude monetary terms, the scale of the 'household economy' has been estimated by quite conventional studies as around 60 per cent of GDP in Australia in 1990.[32] This would suggest that the scale of economic inequalities in the private sphere is by no means 'severely limited'. However, even if Connell's assertions regarding the scale of inequity in the household economy were justifiable, there still seems to me no basis for rejecting the possibility of an economic perspective from the point of view of the household/marriage. Relatedly, I can see no point in ignoring approaches which explore what labour organisation and expropria-tion might look like if economics were not perceived as a singular totality such as patriarchal capitalism. This is exactly what Delphy's work allows: it offers a 'household' perspective, the notion of a distinct sexual economy which examines connections between class and sex after an exploration of specificity in that economy and does not assume harmony at all times between class and sexual systems. She is clearly cautious in her use of Marxism, while recognising a debt to its methodology rather than its content, and does not equate the market with 'the economy'. In many ways, therefore, Delphy's approach seems to be well fitted to the constitution of a sexual epistemology of economics.

Ferguson's account of multisystems (multiple materialities?) could be said to expand the analytical scope of Delphy's essentially dual model and enables a recognition of concerns regarding reductionism or overly unified analyses in relation to class, sex and race.[33] However, apart from this extension, Ferguson's work shares many features with that of Delphy. In both approaches capitalism is *not* indistinguishable from or a form of patriarchy/sex relations, as it is in many of the writings so far discussed. Moreover Ferguson, like Delphy, elaborates the specific character of the 'household economy' as having a particular logic. While she does not centre upon an institutional analysis for investigating the patriarchal sexual division of labour and men's control over women's labour in modern Western societies, Ferguson enunciates a 'sex/affective production' system which bears a resemblance to my earlier term 'emotional economy' and is certainly quite compatible with Delphy's discussion of the organisation of labour in marriage/motherhood.[34] Ferguson's conceptualisation of materiality is more inclusive of emotional, libidinal, sensuous and bodily aspects of women's activities and less preoccupied with an 'objective' approach towards labour than Delphy's (an issue which will be examined in more detail later). This is partly a consequence of Ferguson's closer attention to mothering, sexual subjectivities and psychoanalytic frameworks, as against Delphy's concentration on structural institutions and the 'work contract' undertaken by women in marriage particularly. Nevertheless, when these are added to Hartmann's dual system model, which gives more attention to the historical development of the modern Western sexual system and the aspect of its development in the public domain, and Hartsock's epistemological investigation of a theory of the extraction and appropriation of women's labour and of women themselves, there would seem to be considerable overlap and complementarity.[35]

The several dual/multiple materiality approaches, as I have argued using the example of Delphy, do largely appear to follow the agenda for a feminist political economy outlined thus far. It could be suggested that none of these approaches by itself is sufficient. The comparative lack of an exploration of the role of the state, even from the standpoint of the household, is an issue. However, my own approach in this book will not extend to coverage of the state. Hence I will in this context merely note this as one potential area of inadequacy in the frameworks examined here. Additionally there is the question of the differing attitudes taken in these dual/multiple materialist feminisms towards the significance of the household/family/private domain in characterisations of the sexual economy. While Delphy explicitly views the household as the 'source' for the structure of a sexual organisation of labour, and Ferguson appears to agree,[36] Hartsock and Hartmann are less clear. Hartsock seems to suggest more the critical importance of this site. She refers to the specific form of sexed labour epitomised by, and mostly drawn from, examples of the particularity of labour in the private sphere. Hartmann wavers between a tendency to see marriage/household/family as a crucial means of enabling control over women's labour in both the private and public domains and a seemingly unprioritised concern to distinguish 'all the social structures' throughout society which contribute to that control.[37] The household is, in this latter scenario, only one site among many.

I have already indicated the ways in which I see a focus on women's private labour as an essential and theoretically/politically advantageous stance for a feminist political economy, but this does not necessitate any final decision on whether the household in modern Western societies should be conceived historically as *the* exclusive originating source of sexual labour relations or simply as the most critical site of these relations. Hartmann's disposition to de-emphasise this site at some points in her analysis must be placed alongside certain aspects of her retention of Marxist theory which are likely to induce her to continue to provide a somewhat reductive account of the significance of sex against that of class, despite her championship of dual materiality.[38] Though Hartmann recognises tensions/disjunctions between class and sex and the sex-blindness of class theory, this perspective is virtually overridden by her belief that Marxism, and its analysis of class epochs, provides 'essential insight into the laws of historical development'.[39] Thus she describes a historical periodisation *built around* 'modes of production' ('patriarchal feudalism', 'patriarchal capitalism' and so on). It is not surprising in this setting to find that she is ambiguous about a household standpoint. Indeed at some points Hartmann's analysis is close to Barrett's historical functionalism in that sex relations (though analytically/hypothetically distinct) are defined in practice/historically by the organisation of class relations.[40] This is scarcely a thorough-going reconsideration of social life from the point of view of the sexual system.

It is in fact very difficult to reconceive the history and social relations of Western societies from the standpoint of women and the sexual order without paying central attention to the household/the private domain. Perhaps the field of economics particularly reveals this problem, since 'materialist' feminists with socialist or radical feminist sympathies seem to have great difficulty in coming to grips with labour organisation in any other terms than those of Marxism, even when they consider that they are merely borrowing Marx's method and are critical of his content. Because of this problem, though I have acknowledged the usefulness of Marxism and think sites other than the household may be viewed as significant in the shaping of the economy of sex relations, it seems to me that feminist writers who do not perceive the household as central are inclined (despite their own intentions) to slide into an analytical dependence on class theory/Marxism and into reductive, pacified proposals concerning the sexual order. Hence these writers miss features of the specificity of the economy of the sexual system. This is not, however, to imply that feminist approaches employing a household focus entirely evade an epistemological reliance on Marxism and the accompanying limits imposed upon examination of specificity. I will again turn to Delphy's work to show that even accounts that fulfil many of the requirements of the agenda for a feminist political economy outlined earlier, and that concentrate on the distinct materiality/economy of the sexual system, continue to manifest many of the difficulties associated with a debt to Marxism which inhibit the development of an epistemological revolution in the field of the study of women, sex relations, and economics.

# 6 Conceiving a sexual epistemology of economics and the question of expropriation

> 'Space–time: the error of the West' . . . I should have listened to Lia. She spoke with the wisdom of life and birth. (Umberto Eco, *Foucault's Pendulum*[1])

> An answer is always a form of death. (John Fowles, *The Magus*[2])

## Homing in: Delphy and Marxian materialism

Delphy, in common with Ferguson, Hartsock and Hartmann, does not simply strategically emphasise the conceptual lens of economics and labour, she *privileges* labour over other social processes. For her, the patriarchal sexual division of labour is not merely a basis for an analysis of women's oppression in modern Western societies. Rather she conceives it in the traditional Marxist sense of the 'material' *base*. This organisation of labour is the *source/cause* of the system of male dominance. In typically Marxian terms Delphy describes labour/economics as the 'real'/material level of social life and thus also equates a privileged status for labour, that is, economic determinacy, with materialism.[3] Similarly, Hartmann's 'materialist' analysis of patriarchy is defined by her view that men's control over women's 'labour power' represents the 'material base' of sexual hierarchy.[4] In Ferguson's analysis she locates a sex/affective 'production system' in which 'production' is the causal base of male dominance and economic determinacy is identified with materialism.[5] Hartsock sees the sexual division of labour as linked with 'the *real* structures of women's oppression' [*emphasis added*]. She considers that '[f]eminist theorists must demand that feminist theorizing be grounded in women's material activity'.[6]

I have suggested before that prioritising labour/economics and the paradigm of economic determinacy are at least problematic for a feminist political economy and that feminist materialisms draw upon these Marxist assumptions at a cost. One of the possible consequences of accepting Marxist methodology is that certain Marxist assumptions about the *content* of labour and materialism also tend to be reasserted. For instance, both Delphy and Hartmann tend to ignore the corporeal and libidinal elements of women's activities. While Delphy notes the omission of sexuality at least in her analysis,[7] her account of labour/materialism remains delimited in part by a reliance on Marx's rather circumscribed conceptual scope with regard to sexually-specific bodies.

Delphy and Hartmann are inclined to take for granted that Marx's account of labour is basically satisfactory with only a few modifications of content required for discussion of women and the sexual system. Thus both writers seem to assume that Marx's account of labour within capitalism can stand.[8] This must be questioned in relation to the body *and* the subject at minimum and also, oddly enough, raises the problem of a degree of theoretical/political separatism insofar as Delphy and Hartmann appear to believe that themes of the public domain/market can remain untouched by reconsideration of labour in the private realm. While Ferguson and Hartsock more fully integrate the body and subjectivity into their frameworks—thus implying a possible reworking of the Marxist account of labour in the arena of class relations—they too remain rather silent about Marx's view of capitalism. Like Delphy and Hartmann, these writers seem basically to accept that a materialist analysis of labour in the arena of *class relations* can be the standard one of orthodox Marxism. On the other hand, it must be noted that since the focus of all four writers is upon sex relations, their comments regarding conceptual frameworks for the class system are likely to be restricted in any case. Because discussion of women and class relations is perceived as the second phase of the political economy developed so far in this book, and for reasons of space, I too have made only some schematic observations about the epistemological implications of this political economy for theories of class relations. Nonetheless this field undoubtedly requires further investigation since it may at minimum alter accounts of women's waged labour.

It is perhaps understandable that feminist writers concerned with dual/multiple materiality have not pursued a detailed critique of Marxism in its own terms, that is in relation to class, but other aspects of their acceptance of the Marxist economic paradigm are more problematic. Delphy prioritises labour and identifies materialism with the 'reality' of economics and economic determinacy. The base/superstructure metaphor is reasserted. But while Delphy herself sees her approach as incomplete and particularly notes that she has not covered sexuality, ideology and psychology (subjectivity),[9] it is evident that she believes that the approach is an adequate account of patriarchal economics. Her acceptance of the Marxist base/superstructure model means that she considers sexuality and subjectivity, for example, as 'superstructural'. Further, she equates economics with a kind of depersonalised, quantifiable, 'objective' element of social life (particularly evident in her account of the 'value' of women's private labour) and with coercion/force alone. Economics is viewed as power in the sense of the narrative of mastery. Yet, as Foucault has at least polemically indicated and many feminist accounts of power suggest, power/economics is not merely repressive. Delphy, in other words, takes from Marxism a sharp distinction between and separation of exploiter/exploited founded upon a spatial perception of 'outside' (external coercion/economics) and 'inside' (internal resistance/psychology). As I have noted earlier, this distinction may be inappropriate for the personalised labour relations of the emotional economy in which, firstly, coercion may be highly intertwined with women's own identity/subjectivity and, secondly, economics is constituted not only out of dull compulsion but from the operations of consent.

### Reassessing Delphy's approach in the light of emotional labour and the 'production' of subjectivities

It is no surprise that Delphy concentrates on women's private labours that are visible and quantifiable and does not attend to emotional labours, which deal in the 'production' of subjectivities. Her reliance on a virtually unaltered version of the Marxist concept, base/superstructure, restricts her capacity to examine the range of activities undertaken by women and the particular character of these activities. For Marx the production of products/objects results in the production of social relations, but in the emotional economy this proposal must be extended. 'Production' of social relations in the latter economy occurs in the deeper, more embedded sense that the 'production' process is crucially directed towards *literally*, directly, producing social relations by producing people.[10] If in capitalism social relations are an essential by-product of the creation of things/services, it could be said that in the emotional economy social relations themselves are not a by-product but a (or even *the*) critical imperative. As Hartsock points out, this difference between the class and sex systems suggests the need for some different conceptual tools in the arena of economic analysis: '[o]ne does not (cannot) produce another human being in anything like the way one produces an object such as a chair'.[11] Even the production of objects such as chairs in the emotional economy is organised rather differently than it is within class relations and remains *defined* by a 'subjective' quality.

I contend in this context that Delphy's use of Marxian method has serious consequences for her account and definition of sexual economics, particularly insofar as she largely associates subjectivity/psychology and sexuality with non-materiality, with a non-economic realm driven by the determinacy of economic force. Though these elements are said to have a material aspect, they are seen in typically Marxian terms as secondarily related to the 'economic base' rather than *of* economics (within the sexual system). Moreover, Delphy's use of the base/superstructure model tends to relegate resistance/action/agency to a place outside of economics, or to a side effect of it, insofar as she seems to perceive 'psychology' as an internal aspect of the larger concept, 'ideology'.[12] Delphy pronounces the current absence of and need for 'a truly materialist psychology' in the field of sex relations, indicating her dissatisfaction with supposedly 'idealist' psychoanalytic frameworks.[13] But since this materialist psychology would rest upon a priority given to activities she defines as economic/material—that is, not including a number of forms of emotional labour—and appears to reduce subjectivity to a psychology which derives its shape from an essentially external 'base', Delphy's understanding of both 'materialism' and 'psychology' may be viewed as open to doubt. The analysis is dependent upon an assumption that Marxist concepts/categories, if not content, will do as well for the sexual system as for class relations. To the degree that I have suggested that straightforward employment of a Marxist epistemological framework is inappropriate for analysis of the specificity of women's private labour and may be questioned even in the arena of labour within capitalism, Delphy's exchange model representing the marriage work contract may require some reworking.

Nevertheless it must be noted that Delphy's difficulties in relation to the emotional economy and the significance of subjectivity, for instance, are as she rightly points out linked to the comparative lack of analyses which explore the emotionality/subjectivity of economics. As Thompson states in relation to class theory, analytical concepts and practical investigative methods in the field of the 'subjective' features of work remain very undeveloped.[14] Since these features are so critical to the emotional economy, the lack of concepts and methods is all the more glaring. Delphy is certainly not alone in failing to arrive at a political economy which transforms understanding of labour and the subject. Any criticism of Delphy's work as derivative and partial must be juxtaposed against her highly evocative use of Marxism, which reveals the marriage contract as built upon an illusion of equal exchange similar to that in wage labour. This approach may be most advantageous to feminism and can be used in conjunction with similar perspectives drawn from feminist writers like MacKinnon who are more sceptical of Marxist methodology.[15] Furthermore, even dual/multiple materialist feminists like Ferguson who concentrate on the emotionality of the economics of the sexual system do not necessarily escape a delimited conception of women's activities in the private sphere. Ferguson, in common with Delphy, also employs an 'exchange model' in dealing with labour relations in this sphere. Her association of emotionality/subjectivity with labour is also inclined to tie the 'exchange' logic of the emotional economy to the mothering/production of children and appears to see nurturance/activation of subjectivities in relation to adult men (or adult women) as ordered by this productive 'goal'.[16] The account may underestimate productive 'goals' not primarily driven by child-rearing or *even* by 'the production of people'. Ferguson's assumption that labour in sex relations may be *equated* with the 'production of people' seems to ignore that production of things which does occur in and which is an aspect of the *emotional* economy. She associates the production of things exclusively with the market economy, thus curiously dividing off cooking food, for example, from the sexual system.[17] Such an approach may understate the benefits which accrue to men from women's labour in that system.

Delphy and Ferguson are both inclined to ignore certain elements of labour in the emotional economy; the former because of her use of the base/superstructure metaphor, and the latter because she identifies the 'production' of subjectivities as equivalent to the economics of sex relations. However, these writers in different ways suggest interconnections between labour and 'psychology' which may be contrasted with psychoanalytic frameworks. Delphy asserts in this setting that such frameworks are insufficiently materialist and is concerned that their overly universal and rather static approaches to psychic/psychological processes are inadequate for a historicised feminist political economy.[18] While I have raised some doubts about Delphy's own comments on these processes, I concur with her in questioning the hegemony of the psychoanalytic model for descriptions of the 'production' of the subject. I question the continuing dominance of Marxism and the 'production paradigm' in feminist materialism on the basis, amongst others, of its limitations in dealing with the emotional/subjective character of women's private labour, but this does not mean an immediate acceptance of the dominant position of psychoanalytic themes which do purport to analyse the subjective arena.

While there is not space here to discuss either problems or potentially useful elements within psychoanalytic approaches, I would suggest briefly that what is required is a theory of the subject which allows for examination of a component of *labour* integral to the formation of the subject, does not assume that the psyche exists at a complete remove from 'work', and is capable of historical interpretation. Two of the most significant feminist approaches in the field, those of Chodorow and Mitchell, tend to revert to dehistoricised accounts when discussing the subject[19] because they remain, despite some level of critique, bound to a Freudian/Lacanian model, and are at least somewhat vague about the relationship between labour and the psyche.[20] Additionally, these approaches tend to stress different elements of the 'production' of the feminine psyche/subject and are hence seen as conflicting. Yet the supposedly different elements might be seen as capable of amalgamation rather than as dichotomous. While Chodorow stresses the symbolic/psychic significance of the Mother and sexual identification, Mitchell places emphasis upon the Law of the Father and object choice. Both approaches could be employed in discussing contestation between the law of phallus and the feminine signifier and hence ambivalence in the ongoing formation of the modern Western feminine subject from childhood through adulthood.[21] Perhaps what is also necessary is to break with the notion of a singular psychoanalytic 'myth' relevant to the subject (just as I have suggested breaking with 'mainstream' and Marxian economyths) in order not only to contextualise this concept historically, but also to recognise the specificity of women's situation.[22] If this is accepted neither Mitchell's version of the Law of the Father (object choice orientation), nor Chodorow's account of the influence of the Mother upon sexual identity (identification orientation)[23], may be sufficient. It is at least possible that different 'myths' may be relevant to femininity/women as against masculinity/men as the hegemony of Freudian/Lacanian theories of the subject is challenged. The omnipresent singularity of the Oedipus complex might be reconceived for examinations of the construction of femininity in modern Western societies.[24]

Freud/Lacan, Mitchell and Chodorow adopt approaches in which *either* the Law of the Father or the Mother shape the subjectivity of *both* sexes. It may be that these comparatively singular alternative accounts are mistaken in assuming that one mythical/symbolic 'path' can describe sexually-specific bodies, sexualities and subjectivities in Western sexual systems. In other words, even if the myth of Oedipus were appropriate for examinations of (Western) masculinity, perhaps it might be useful to consider whether the construction of femininity occurs under a different mythical regime. I have suggested that sexual positioning might be characterised in modern Western cultures as a process of contestation between the symbolic orders of the Mother and Father with regard to femininity. If this is so, the myth of Persephone may be more appropriate for analysis of femininity than either Freud/Lacan's or Chodorow's versions of the psychoanalytic paradigm.[25]

The myth of Persephone depicts her as the daughter of Zeus (the Father of all) and Demeter (Earth Mother/Virgin). Persephone's mother keeps her beautiful daughter safe by shutting her away on an island. But Zeus accepts the request of his brother, Hades (another incarnation of Zeus the Father), to marry the girl. In the face of Demeter's known objections, Zeus plots to trick his daughter into marriage. One day when Persephone is alone she sees a narcissus flower (representing self-absorption and hence disobedience to/separation from the mother). At this point Hades rises in his chariot from the underworld and abducts Persephone against her will. Demeter discovers Zeus's betrayal and eventually, by threatening Zeus with the loss of her fertile favours, discovers the whereabouts of her daughter. She contrives to win Persephone back. However Persephone has partaken of fruit in Hades's domain and thus Hades claims his rights as a husband. A 'compromise' is arranged by Zeus–Hades. Zeus decrees that Persephone must spend part of her time with Hades (as the queen of the dead) and the other part with her mother. The earthly domain of Demeter becomes joyous and fertile when Persephone is with her mother and passes into winter/barrenness when she departs.

In this myth the mother is both protector and warden and the father betrays mother and daughter, collaborating in the (incestuous) rape of the latter, which is masked as seduction/marriage. The daughter is tempted by the pleasure (and status?) of heterosexuality and indeed appears at least to accept her fate, since she is always depicted as acting in concert with her husband. Yet the myth also shows the bondage, sadness, and debasement of heterosexuality in patriarchal social relations, as well as the daughter's continuing bond with her mother. She 'dies' in a man's arms as his wife, yet is rejuvenated by her mother's ongoing (possessive) love. It is an account of the divided, unstable character of (heterosexual) femininity: a path of continuing contestation punctuated by the regular reunion of Maiden and Mother and their sorrowful but inevitable partings. Interestingly, it is not the heterosexual bond which gives rise to creative/fertile 'production', but the plaiting of two aspects of the feminine across (against?) the braid of patriarchal power. This schematic outline is simply offered as a possible mode for reconceiving aspects of the theory of the subject with an eye to the specificity of femininity, women's experience and women's labours. Clearly this field requires considerably more analysis than can be attempted here.

Finally, in common with Foucault's approach, the historicisation of the subject might involve a more conditional stance in relation to the psychoanalytic equation of sexuality with subjectivity. While sexuality may be identical with the constitution of the subject in the recent or even distant history of the West, this assumption must be open to debate, especially if theories of the subject are considered cross-culturally. Similarly, the psychoanalytic inclination to register an equivalence between the psyche and the subject may be questioned. A feminist political economy requires analysis of conscious as well as unconscious processes and must ponder whether the unconscious can indeed be taken as a kind of determinate 'base' for the complexity of mechanisms related to subject formation.[26]

## Delphy, institutional frameworks, and the critical problem of expropriation—the limits of indirect methodologies

Much of this discussion of the reconception of theories of labour and the subject would not be at odds with Delphy's proposals, though it would involve an exploration of aspects of the character of women's labour in the private sphere and the definition of economics which would have implications for Delphy's use of Marxist methodological procedures such as the base/superstructure model. Delphy can, additionally, be credited with offering a pioneering analysis of processes of *expropriation* in relation to women, sex relations and the household. Her framework does provide a politicised account of the specificities of the domestic economy and a series of conceptual tools for a detailed investigation of power in the distinct labour relations of the sexual system. Delphy's work in this area arose at a time when feminist writings on women's *labour* were primarily concerned with waged labour and there were almost no extended discussions of the nature of expropriation in sex relations.[27] This largely remains the case. Not only has there been comparatively little research on the household economy, but there are no clear indices for valuing work in the home, and no clear agreement about the indices which are used, let alone adequate analysis of the connection between such indices and expropriation.[28] These problems are exacerbated by difficulties in drawing distinctions between 'work' and 'non-work' (exploitative and non-exploitative labour)[29] and underdeveloped tools for considering 'subjective' elements of labour. Such definitional questions suggest awkward issues for all forms of labour studies but they are particularly contentious and central in the area of household work. Delphy does at least provide a procedure for investigating household labour and valuing that labour in a framework describing expropriation.

Delphy proposes a methodology oriented towards institutions, particularly the marriage institution, and argues that expropriation can be estimated in this institution. For her women's private labour is gratuitous, since as long as women perform this labour for people to whom they are not related or married it acquires exchange value and is remunerated.[30] The valuelessness of domestic work performed by women is hence related to a work contract in which husbands appropriate all work done in the marriage institution, especially that of their wives, and moreover can sell it in waged work as their own if their wives contribute to this wage labour.[31] In Delphy's view, modern Western women's private labour is done for nothing. It is a profoundly disturbing account of exploitation in labour and the benefits of this for men. The stark picture she provides could undoubtedly propel a feminist political program for change and this is of some significance in the evaluation of feminist political economies.

Other feminist writers have outlined similar institutional approaches but there has been considerable debate over which institution is central to the sexual system, partly as a result of differing levels of focus upon labour relations and the private sphere. Delphy herself appears to give some credence to both the institutions of marriage and motherhood. In similar fashion, Dworkin refers to farming (motherhood) and brothel (sexuality) models of patriarchal sexual organisation.[32] Feminists such as Rich and MacKinnon prefer to see the institution of heterosexuality as critical, while O'Brien, Al-Hibri and Kittay draw upon modes of organisation largely framed by the centrality of child-rearing and motherhood.[33] Ferguson, in common with the last three writers, focuses on the production of people, but she does also pay attention to the institution of compulsory heterosexuality.[34]

Perhaps here, as before, a multiple methodology for analysis of labour and expropriation might be useful. I have noted in Chapter 2 that recognition of the variety of activities performed by women in the private sphere and of a possible range of criteria for analysis/measurement of these activities requires a kind of multiple analytical matrix for the study of domestic labour. I have also suggested the advantages of a dual/multiple materialist feminism which conceives at least two specific modes of economic organisation. In keeping with this approach I concur with Jaggar's point that, while patriarchal sexual order *may* have existed throughout known history, it is unlikely to have existed in an unchanging form and thus there may be problems in attributing sexual hierarchy or expropriative sexual labour relations to a single source[35], or for that matter locating a single institutional framework. On the other hand, it may be that discussion of the domestic economy and labour in sex relations would be better dealt with under one institutional focus than under another. This question has barely been debated and requires substantially more elucidation than is available in the feminist analyses mentioned. Comparison and evaluation of the differential potentialities of the several institutional models is well overdue, but at present Delphy's focus on marriage certainly offers one of the clearest and most detailed frameworks for consideration of the specific form of exploitation in sex relations.

This clarity is especially evident, as I have already suggested, in the area of defining the nature of expropriation and the value of women's work in the sexual system. Most other feminist writers remain somewhat vague on this difficult subject. O'Brien states that men steal the value and products of women's labour, but it is not clear *how* this is done. Like Al-Hibri and Kittay, she tends to refer broadly to men's control over women's (pro)-creativity. The emphasis is more upon the effects or types of this control than on how it occurs.[36] Ferguson is rather more specific and outlines a definition of expropriation in which overall control over women's labour is related to unequal input, unequal control over the labour process, unequal access to outputs, and unequal benefits and satisfactions.[37] Moreover she refers, as does Delphy, to questions of control that go beyond simple decision-making authority over persons and things.[38] Cohen has argued, in her critique of Marxist labour process debates, that emphasis on the notion of 'control' in the labour process may lead to a focus merely on the political aim of limited forms of decision-making power over aspects of job control and ignore issues of 'ownership'.[39] Ferguson and Delphy exhibit a concern to enunciate a pattern of organisation that is not restricted to an analysis of control over the job of household labour, since after all it could be said that women have a considerable degree of leeway in this regard. They note the ways in which expropriation of women's labour is tied to a set of labour relations. This corresponds to Cohen's concern with ownership in the sense of describing a structure of legiti-mate/sanctioned possession justified by rules or norms.[40]

Though Ferguson and Delphy pay little attention to access to resources and technological 'means' in the analysis of posses-sion/ownership, their approaches do enable exploration of appropriation in terms not only of women's labour and its 'products' but of women themselves.[41] Such accounts of control in the labour process *and* of ownership of women's labour/women can enunciate a mode of labour relations which stretches out from the household to institutions like the state and can incorporate in the analytical framework the 'poverty of protection' available to women within the domestic economy in areas like the law.[42] The 'environment of patriarchal exploitation' is thus perceived in a broader framework than that of achieving simple control over the job of domestic labourer.[43]

Delphy and Ferguson attempt accounts of the mechanisms of exploitation in the domestic economy which in different, complemen-tary ways elaborate the complex fabric of control/ownership –possession with regard to women's labour in the sexual system.

Therefore, while elements of both approaches could be said to require alteration and/or addition, they do suggest a basis for a sexual epistemology of economics. Delphy, however, more than Ferguson, proposes a model for estimating/analysing the extent of expropriation and the value of women's work which might be useful in considering variations in the intensity of 'exploitation' over time.[44] The model adopts an aggregate view of women's activities in the household and can therefore only be employed as one broadly based methodology amongst many. Furthermore, Delphy's analytical technique for registering appropriation is founded in market measures; that is, she notes women's lack of remuneration in private labour against that which they would earn if they undertook this labour in the capitalist marketplace and also men's appropriation of women's private contribution to men's 'exchange value' activities. Private expropriation is judged by its market value or its impact in the market economy: in both cases the measure of expropriation in marriage is found *outside* the marriage institution and is ultimately monetary. Delphy's reworking of the Marxist labour theory of value remains derivative. This is not to suggest that aspects of expropriation in sex relations cannot be discussed in terms of women's unequal access to monetary rewards and their contribution to men's monetary rewards, but the indices, Delphy relies upon are derived from an economy external to the one she describes and from an economy which is substantially different in character and logic from that of the marriage institution, a difference which her own analysis shows.

In fact Delphy offers measures which in some ways simply involve a more politicised reading of indices employed in household work studies. Her description of women's lack of remuneration is very like the 'market cost approach' which summarises the market cost of separate services undertaken by women and is related to the 'opportunity cost approach' which estimates women's potential earnings. Delphy's delineation of women's contribution to men's waged work is rather more interesting but in many ways represents an element of the 'replacement cost approach' which is intended to estimate the cost of replacing 'the home manager'. To the extent that Delphy outlines the continuing appropriation of women's labour after the dissolution of marriage with regard to mothering responsibilities, she moves beyond straightforward usage of market/monetary measures. Her analysis, however, parallels market-based discussion in household work studies which shows the drop in women's incomes, and a corresponding rise in men's, after divorce, indicating both the inequitable effects of child-rearing upon women before and after marriage and the unequal costs of marriage registered in differential divorce outcomes.[45] What is problematic about all the indices noted so far is that they register expropriation in primarily market/monetary terms and/or after divorce. In both cases the indices represent expropriation through locations 'outside' of its direct practice in the household. There remains an overall failure to conceive direct measures/forms of analysis.

## Suggestions regarding a direct approach: a typology of measures/indices

The few seemingly 'inside'/direct measures that are proposed are promulgated within household work studies but these usually refer to *time* usage or *output* (amounts of visible services/products). As noted previously, time itself is most often conceptualised from a market standpoint and thus, though it can be employed to consider certain activities in the private sphere, is of uncertain use in relation to emotional labours which can be said to epitomise the emotional economy. For example, does more or less expropriation take place if sexual activity takes more or less time? Similarly, what Umberto Eco has described as the 'therapy of the word'[46] cannot be adequately evaluated by measuring the time it occupies since time does not fully reveal expropriation nor degrees of expropriation, and in any event this activity is ongoing. Quantitative measures in such circumstances must give way to discussions of qualitative meaning, or at least time must be understood in the sense of meaning, as a dimension which varies (lengthens/shortens) with the position of the observer rather than as a constant. Output is an equally limited measure for emotional labour and indeed its usage is commonly related only to activities which could be performed in the marketplace. This reveals another weakness in both indices, which are employed by household work analysts and by Delphy. Such indices are associated with a concentration upon monetary value in particular and display a tendency to ignore labours that could only with difficulty be translated into the market economy. Essentially these frameworks concentrate upon visible so-called 'productive' activities in the household/marriage and largely evade the question of how to value women's emotional labour and expropriation associated with this work. If what is distinctive about women's private labour and the economic organisation in sex relations is precisely the logic of emotionality, of deeply personalised labour relations tied to love, affection, sexuality and relational bonds, then it is of some concern to find that the indices applied to women's work hardly touch upon the crucial area of direct investigation of emotional/subjective labour.

Though it can be said that the household economy as a whole is under-researched, emotional labour appears in this context as the equivalent of the Lacanian concept of 'lack'. It represents a black

hole, a deeply invisible feature of women's experience.[47] The unqualified specificity of emotional labour in the private sphere makes it
particularly resistant to derivative methodologies. Perhaps this is a
partial explanation for evasion of its presence. To the degree that
emotional labour is recognised, it is generally perceived under the
rubric of childcare (and then only in limited terms). This is paradoxically the more politically acceptable face of emotional exploitation.
It is after all far more difficult to acknowledge the inequalities of
love and conjugality that go to the centre of relations between men
and women. It is indeed extremely disquieting to reassess aspects of
those relations which are fondly characterised as natural (irrational,
eternal, essential) and voluntary feelings as a part of an historically
formed system of exploitation of labour. Nevertheless, analysis of the
emotional economy requires not merely a deconstruction of notions
of 'the economy' and 'work', but of emotionality/subjectivity. Indices
of expropriation must accordingly be far more varied and must
consider the emotional economy in its own terms as well as those
of the market.

Three possible 'orders' of measurement/analysis might be considered which indicate variable degrees of application of indices like
time and money.

*Direct* measures could be elaborated that particularly locate
qualitative estimates of meaning, including patterns of demarcation;
transferability of certain activities; patterns of negotiation; definitions
of and 'levels' of satisfaction; distinctions between responsibility for
and direct 'ownership' of jobs; ethics, notions of justice and 'proper'
behaviour; hierarchies of tasks that affect which takes precedence
and at what times; perceived intensity of different tasks; the status
of household work overall and for different activities; ownership/pos-
session of persons and things; control over the labour process in
general and in specific areas, et cetera.[48] Quantitative measures like
time usage can be regarded as 'partial' indices relevant to some
activities or elements of these activities under the heading of direct
measures, whereas money cannot be included here.

*Proximate* (or near) measures might also be employed to study
the emotional economy. An example of this order of estimation of
value/expropriation could be evidence of the advantages of marriage
for men against women that occur in figures showing that while men
live longer if they marry, women who marry have reduced lifespans.[49]
Time but not monetary measures could be used under this heading.

*Proxy* measures may be employed which include 'substitute'
indices like time and money. These indices are likely to be derived
from economic frameworks outside of the emotional economy, such
as the marketplace.

There are strong political reasons for advancing more sophisticated valuations of household work and expropriation within it. Divorce courts are increasingly attempting to make judgments about women's contribution to marriages and unless more adequate measures are found in this arena, women will continue to be disadvantaged.[50] Currently in Australia women's contribution tends to be closely tied to 'needs-based' claims concerning their unequal responsibility for childcare.[51] The dimensions of the labours linked to childcare have as yet been narrowly interpreted because of inadequacies in the valuation of household work.[52] Moreover, estimations of this work have largely evaded activities such as care beyond those associated with children. This evasion exists side by side with the maintenance of legal provisions in the state of Victoria, the Northern Territory and the Australian Capital Territory which allow men to claim compensation for loss of services and companionship (loss of consortium). This compensation covers not only instances where a wife is ill but even those where she is 'moody'.[53] Legal provisions regarding loss of consortium were once available to men throughout Australia, but were not applicable to women since 'they were not regarded as having a property right to their husbands' services[54].' The paradox of a general lack of recognition of the value of women's unpaid labour as against the historical and current existence of laws allowing action to compensate individual men for loss of domestic work has prompted Marcia Neave to comment that '[t]o the cynical eye this may suggest that men recognize the value of domestic labour only when they are deprived of it'.[55] Additionally, a 1990 challenge to the South Australian 'rape-within-marriage' law mounted by George Romeyko depended on an interpretation of the Federal Family Law Act as encompassing the *right* of one spouse to be 'sexually comforted' by the other, rendering rape in marriage an invalid offence and considering the obligation of sexual comfort a part of marriage.[56] Though the High Court ruling against Romeyko's challenge to this South Australian law is reassuring, Seddon has pointed out that the Family Law Act is ambiguous and does appear to imply that marriage involves the consent of a wife to provide sexual services.[57]

Whether or not these legal provisions and legal debates continue or are affected by legal reform, it is clear that there remain some grounds for recognition of various emotional/sexual labours undertaken by women as part of the rights of men, but there is little in the way of an alternative vision of these labours as representing expropriation. To the extent that domestic labour is valued, it is recognised in divorce, largely after expropriation has already taken place. There is presently no established means for evaluating women's private labour and ameliorating/avoiding/preventing the expropriation of that labour *before* it happens. Family Law is based on the notion of an equal partnership within marriage, despite the fact that this is contradicted by its recognition of inequalities in the 'costs' of marriage during divorce proceedings.[58] The organisation of direct expropriation remains untouched. This makes the issue of developing indices appropriate to women's private labour, and emotional labour in particular, all the more pressing. However, there are other political and theoretical reasons why analysis/measurement of the gamut of women's household work and expropriation in this should be made visible. At present there is no system of arbitration of labour relations worldwide that is concerned with the private sphere and with the potentially differing interests of men and women in relation to private labour. This is despite the fact that organisations and legal procedures devoted to the arbitration of the differential interests of capital and 'labour' exist in Australia and elsewhere. There is no equivalent of industrial relations law that exists to consider or defend women's position in the labour relations of the sexual system beyond certain very restricted and largely 'after the fact' features of Australian Family Law. This problem at least becomes an area for debate in the process of developing indices associated with women's private labour.

Furthermore, conceptions of 'development' and analysis of possible intensification of exploitation in the Third World could begin to look very different if one no longer, as Mies is inclined to do, simply described unpaid domestic labour in class terms as a hidden base for the extended reproduction of capital.[59] 'Development' may then be viewed in relation to the (partial) imposition of Western forms of specifically patriarchal labour relations which may reveal other dimensions to imperialism than are available through a primarily market-oriented framework. Finally, the growing significance of environmental concerns may be conceived in other ways if effects on women's private labour are explored. Until now the crucial focus of these concerns has been upon the impact on men's jobs in areas like the forestry industry, but this does not take into account the potential intensification of women's labour that might be required to achieve many environmental ends and hence fails to examine potential resistance to or problems in achieving those ends.[60] This brief discussion enunciates only a few of the possibilities of conceiving a feminist economics but clearly, to the degree that the specificity of that economics is acknowledged, there are far-reaching implications for epistemological frameworks, political aims and policy objectives.

It seems that not only are a feminist 'materialism' and a feminist economics possible but that these engender a vision of social life through women's eyes, through a lens that offers an irreplaceable view simply not within the scope of dominant concepts and categories. This vision is no more available through Marx's analysis of economics than it is in the practical statistical surveys developed by Ironmonger. Feminism must develop a detailed account of the sexual epistemology of economics to place against the hegemony of such approaches and the language and perspectives of 'economics': indeed it cannot afford to do otherwise. Women's labour with all its complexity and unique qualities will continue to be expropriated while it remains a shadowy backdrop as yet only partially conceived. Though I have suggested that no single epistemological framework, methodology or terminology can be sufficient for the conceptual task—to paraphrase the quotation from Fowles which heads this chapter, an answer in any analysis of social life suggests a deadening paralysis of thought and practice—this book attends to a combination of approaches, thereby attempting to embody the shadow and transform it into a visible presence.

# 7  Conclusion: charting an/other direction

> Women have been largely absent not only as economic
> researchers but also as the subjects of economic study . . .
> The first edition of Paul Samuelson's *Economics* (1948) had
> only two references to 'females' and none to 'women', both
> included in a segment on 'minorities'. Even today, women
> and families remain strangely absent from many 'general'
> discussions of economic matters . . . Certain activities and
> experiences that are historically of greater concern to
> women than to men have all too frequently been neglected.
> Further, even when economists have attempted to
> understand phenomena from such traditionally feminine
> realms as the home and family, the results are often judged
> as unsatisfactory by feminists who believe that the analysis
> of women's experiences is inadequate or even biased.
> (M. Ferber and J. Nelson, *Beyond Economic Man*[1])

I referred at the beginning of this book to Western feminist debates
around 1980 in the field of economics and labour, in particular to
debates between Barrett and Delphy, who considered the question of
whether a 'materialist' feminism was possible.[2] I noted that
exchanges between these writers, among others, raised related issues
such as whether a feminist economics could be elaborated. This book
is intended to demonstrate that it is indeed possible to conceive a
'materialist' feminism, and relatedly a feminist economics, dealing
with modern Western societies. However, I have also suggested that
such a conceptualisation requires a redefinition of 'the economic'.
That redefinition involves placing under scrutiny the concepts, categories and methods typically employed in economics. This can be
undertaken by paying attention to the specificity of modern Western
women's work, especially their private labour.

In the process of reconsidering economic frameworks, I have
concentrated on those approaches which have held sway in the field
of *feminist* accounts of economics and labour. Hence, my focus upon

Marxism and the more 'centrist'[3] or 'mainstream' perspective generally found in household work studies. I contend that such economic paradigms are inclined explicitly or implicitly to privilege the market and market measures and consequently ignore/marginalise the specific features of women's work epitomised in the domestic economy. While I do not suggest that the methodological procedure of drawing comparisons between market labour and that which occurs in the private sphere of Western societies be abandoned, my central argument is that exclusive or privileged use of this procedure delivers a seriously inadequate analysis of the latter and is therefore undertaken at a cost. The cost is high, for this use of market-derived paradigms results in a continuing lack of visibility for a range of women's labours, particularly those related to emotional labour/care. Since household work studies, for example, have acknowledged emotionality/care as crucial to the 'logic' of operation of the deeply personalised, emotion-laden domestic economy, this evasion of the implications of the *difference* between market and household labours is both ironic and of considerable concern. The cost indeed lies in ignoring the particular form of labour characteristic within modern Western sex relations and failing to consider private labour *in its own terms*.

I have asserted in this context that insofar as Marxism and household work studies privilege market labour, and hence perceive the household economy through the lens of an economic system which is external to and different from it, they are bound by 'sexual economyths'—that is, by economic assumptions which exclude/marginalise women's experiences. These approaches give priority to the field of the market and offer a lesser, comparatively invisible, positioning to that sphere of labour which characterises women's work, that is, the household economy. The economic epistemology of Marxism and household work studies may be viewed in this sense as replicating a masculinist hierarchy of meanings and value which 'distort[s] our understanding of all social life by ignoring the ways in which women and gender shape [that] social life'.[4] I have consequently proposed, in line with Harding's demand for 'a revolution in epistemology',[5] that what is required is an economics which takes women as the *subjects* of the analysis and its point of departure from the specificity of women's labour. Clearly the feminist political economy outlined is no more a total picture of economic processes than those approaches founded in male and market activities. However, if feminism has a concern to explore sexual forms of power and expropriation, one important method of doing this is to take women/sex relations and the household as the starting point for the study of labour and economic forms. This sexual epistemology

of economics challenges sexual economyths and places power relations between the sexes at centre stage.

Furthermore, feminist economic analyses are themselves frequently subject to many of the same problems as those raised in
relation to Marxist perspectives and household work studies. The
critique of sexual economyths underlying the latter approaches
encourages reassessment of existing feminist paradigms. In this reassessment, my outline of an alternative economic epistemology
suggests a direction more along the general lines of Delphy's viewpoint than that of Barrett. Nevertheless, the reconsideration of
feminist accounts of labour and economics indicates a common
propensity to employ epistemologies which are market derived, often
as a result of the employment of Marxist categories, concepts and
methods. In this context I suggest that the development of a sexual
epistemology of economics founded in the specificity of women's
labour and the household economy, in their difference from market
labour and the market economy, involves an alternative not only to
commonly employed 'mainstream' and Marxist economic frameworks but also an alternative to much contemporary feminist
economic analysis.


## Why bother considering an alternative?

Judith Allen, partly drawing on the outlook of Catherine
MacKinnon, has stated that feminist theory generally 'has been a
tolerant fellow traveller, along routes dictated by the theoretical needs
of others, for quite long enough!'[6] In the realm of economics I can
only repeat this claim 'with bells on'. Feminist theorists, policy
makers and service providers must in my view look beyond market
models, not only because overtly celebratory accounts of the market
may be criticised for an overly simple account of socioeconomic
processes in the public sphere and for an individualistic orientation
that underestimates the impact of class positioning—which are
common social liberal/social democrat/socialist concerns,[7] but additionally because market-derived models, whether of a 'mainstream'
or Marxist variety, are based in masculinist assumptions regarding
the significance and character of women's labour. In other words,
highlighting the requirement for feminists to be sceptical about the
advantages of clinging to the coat-tails of epistemologies which were
not founded in feminist terms of reference is of particular importance
in economics. In this setting, many feminist writers in the field of
economics have drawn attention in various ways to the need for
altered premises. As I noted in the introductory chapter, Eva Cox,
in her work as a feminist activist within the field of economic policy,

has expressed dismay at the lack of an alternative economic vision which might be placed against the dominance of economic rationalist approaches. Cox calls for a viewpoint which is not tied to the individualism of the market but which expresses the 'social imperative of mutuality'. She is understandably cautious about any attempt, in response to the dominance of market perspectives, to romanticise the private realm.[8] All the same, it is clear that she associates 'altruism'/'mutuality' with a movement away from an exclusive focus on the market towards the potentialities of perspectives which explore the specificity of the private sphere and, in particular, its difference from the atomised 'greed' of the market. It would seem that the approach taken in this book may offer at least in some broad respects a contribution to the alternative Cox seeks.

But of course there remain numerous problems in proposing guidelines for such a contribution. Rebecca Blank, from a neoclassical economics background, expresses impatience with feminist approaches to economics that criticise existing economic paradigms but do not themselves provide 'a clear model' to follow. Indeed, she says of the essays she was invited to discuss in *Beyond Economic Man*, that 'a reader who hopes to glean a clear model of feminist economics from this book will come away disappointed.'[9] However, she does note that

> it is surely far too strong a *standard* to ask for a fully developed model at this stage of analysis. New paradigms are developed over time; they gain structure and complexity as more people become interested in them. Because scholars have just begun to apply feminist theory to economics, it seems somewhat churlish to complain that the full model isn't clear yet.[10] [*emphasis added*]

The guidelines I have outlined may be subjected to the same (in this case reasonably sympathetic) critique. This book does not provide a completed model but rather some hopefully useful parameters which can potentially provide a direction in the relatively new and uncharted waters of a feminist economics. On the other hand, the clarity that Blank considers feminist analyses ought to aspire to achieve appears to reside in the mathematical modelling and associated 'empirical results' that she sees as constituting critical 'strengths' of neoclassical economics. Blank constructs this methodology as the 'standard' against which a feminist economics should be assessed and relatedly argues that feminist analyses of economics must move towards this standard to 'gain wider interest and respect from the [economics] profession.' She essentially judges the essays in *Beyond Economic Man* to be lacking in rigour. Blank charitably attributes this to the as yet early stage of development of the field. Even so, she perceives feminist approaches to economics as largely limited to

vague principles which do not create 'a clear sense' of secure (even
if limited) knowledge.[11] By contrast, neoclassical economics conveys
this security:

> [m]athematical formalism is one way (though not the only way) to
> produce theoretical precision. As such, the high degree of
> mathematical rigor in economics creates a clear sense of what we
> know and what we don't know, at least within certain limits: We
> [sic] know which problems have been 'solved' and which have not.[12]

Blank's formalised model of 'the economy' brings to mind a particular kind of physical 'model', a shining metallic object with clear
cut sheer facets and no fuzzy edges. This is not to dismiss her
perspective but rather to recognise that, while Blank is careful to
acknowledge that such a model is an abstraction and therefore
'partial', the sharply defined contours of the model make it especially
amenable to a conception of knowledge as discrete 'facts'. These
facts at least appear definite and can be collected by available
computational techniques, manipulated and used. I do not dispute
that there may well be certain kinds of advantages in this vision of
'the economy' as capable of ready translation into 'tractable mathematical forms'.[13] It does provide a way to think about complex
processes and come up with 'solutions'. Moreover, it engenders a
sense of capacity to manipulate or control those processes. Nevertheless, even if the benefits of using quantitative modelling were
entirely accepted, there remain two crucial issues of relevance to
Blank's discussion of feminist economics.

Firstly, assuming—and this is obviously contentious—that the
neoclassical model Blank employs describes reasonably adequately
the forms of the market economy from which it is *derived*, does this
model describe the household economy to which it has merely been
*applied* as a recent afterthought? This is not just a question of the
necessarily partial quality of abstract frameworks. What if the neoclassical model ignores not only the confusions of complexity in its
abstracted formalism but fundamental characteristics of the household economy? In this context, as against the image of a distinct
metallic object with clearly definable features that can be subjected
to tidy measurement, women's private labours appear as an opaque
miasma or shadowy sea with no clear beginning or end. This sea is
by no means formless or chaotic: it may be said to display certain
currents, tides, turbulent crosscurrents and repetitive wave-like con-
figurations. Such characteristics are likely to be amenable to some
use of mathematical analysis. But is the emotionality of the household—its critical attribute—productive of 'theoretical precision' along
the lines Blank takes as the appropriate aim of a feminist economics?
The second question, in other words, is 'would "mathematical

rigour" (numerical measurement) give rise to a clear sense of what we know and what we don't know, or how to "solve" problems in the field of the household economy?' I argue that the household economy differs in many decisive respects from that of the market and that consequently application of identical measures to both— measures drawn from the characteristics of the latter—must be questioned. In particular, I assert that the privileged or exclusive use of market measures is a seriously flawed methodology. Hence Blank's perception that feminist approaches to economics fail to provide a 'fully developed model' largely on the basis of their inattention to mathematical precision may require further examination.

On the one side I believe Blank offers a fair comment on what is, as she herself points out, an emerging form of feminist scholarship.[14] This book is concerned with offering some general epistemological guidelines. It is not in itself an example of empirical research nor does it provide a detailed methodology for empirical studies, though it is hoped that the framework will assist in the shaping of such research and research methodologies. However, because feminist economics is an emerging field, I consider it highly appropriate to outline a possible epistemological framework and am inclined to believe that to undertake empirical research or even develop detailed practical methodologies in the absence of clarification of appropriate epistemological paradigms is a difficult task indeed. As Blank comments, mathematical rigour is not the only means to theoretical precision. I would argue that the construction of epistemological frameworks suited to the field of women's labour and the household economy is a significant, perhaps crucial, aid in the process of developing theoretical precision. If the means to such precision were taken to be the elaboration of a 'clear [mathematical] model' in the terms Blank outlines, it is likely that even this conception of precision would be greatly advanced by clarification of the epistemological premises which underlie such mathematical modelling. Thus, while Blank has some grounds for pointing out the limits of existing feminist approaches to economics and some basis for her impatience with what she views as a lack of empirical orientation, the concern of feminist commentators with 'general theoretical statements of principle' is not inevitably to be associated with imprecision. In fact, close attention to frameworks that might be employed in the field of feminist economics is a critical means to the precision Blank apparently seeks, even if the mathematical *dimension* of precision which she regards highly is (and may remain) undeveloped.

Such an appraisal of Blank's comments is intended to point out that while this book is generalised in approach—in common, it seems, with many other feminist analyses of economics—the refinement of theoretical principles is a necessary and important part of the constitution of a feminist economics. Blank, correctly I believe, notes some of the limits of an overemphasis on outlining epistemological frameworks, but at the same time underestimates the value of these frameworks in a developing field, particularly one which may differ from existing arenas of knowledge and require new ways of thinking. This point gains weight if Blank's observations regarding the problems of an unbalanced focus on 'general theoretical statements of principle' are viewed as only appropriate to the single volume she is considering.[15] Beechey, whose claims to a wide-ranging knowledge of feminist literature relevant to economics are beyond dispute, asserts that 'with a few exceptions, the field of work has become a rather atheoretical area of feminist intellectual enquiry'.[16] There would seem to be considerable room for more precision in feminist epistemological frameworks dealing with economics and labour.

The other side of this appraisal involves a somewhat more critical view of Blank's suggestions. She appears to assume that, whatever the arena of a feminist economics might look like, it will necessarily be amenable to mathematical modelling. Since economic processes are presumed to inevitably be capable of quantitative measurement, Blank concludes that if feminist scholars wish to be taken seriously in the economics profession they must undertake the necessary measurements. However, I believe there may be serious limits to the methodology of measurement with regard to labours within sex relations. In this setting, commonly employed forms of measurement which are deemed appropriate in the market may be used in conjunction with other forms of analysis in research on the household economy, but this book has attempted to show the difficulties of ignoring the restricted scope of such measures and of failing to devote attention to modes of analysis which are drawn from the specificity of the household economy. If this argument is accepted, it would seem, according to Blank, that feminist scholars in the area of economics are in a double bind. Either they must accept the sexual economyths of market-based 'mainstream' or Marxist approaches and ignore the particularities of women's labour or, as Judith Allen suggests, refuse fellow-traveller status and forgo the 'interest and respect' of the economics profession.[17] It is to be hoped that these are not the only choices available, but Blank's account of the outcomes of challenging existing frameworks may be correct. What is of concern is that she does not for one moment appear to imagine that if feminist economists should find quantitative measurement ill–suited to their task, then the economics profession might concede that its 'rigour' could be viewed from the perspective of that task as lacking precision. Blank assumes that the imprecision is largely one way. Though she acknowledges that feminist economists might add something to economic modelling, she does not envision an/other starting point from which it may be said, for instance, that economic modelling *might* add something to a feminist economics. I have noted in Chapter 3 that this unidirectional outlook is characteristic of the sexual economyths of 'mainstream' and Marxist economics, that is, women and their specific experiences are typically construed as dependent categories of analysis.[18]

In the light of this critique, Blank's perception of feminist approaches to economics as failing to provide a 'fully developed model' may be regarded as a partisan judgment. Her perspective carries the stamp of phallocentric epistemologies insofar as it is reliant upon a market 'standard' which is likely to make assessment of feminist approaches as callow and insubstantial (compared with the supposed maturity of conventional economics) a foregone con-clusion, however 'precise' those approaches might become in terms of the particularities of women's labour. Though I would not argue with the description of feminist economics as embryonic, I consider existing feminist frameworks as underdeveloped in ways that Blank does not and, furthermore, there seem to me grounds for questioning an analysis that appears to refuse to grant credence to the alternative vision of economics which these frameworks might glimpse and over time bring into view.

However, Blank does outline a further basis for admonishing feminist accounts of economics which I think gives rise to a legitimate and even-handed recommendation.

> The feminist economics literature is too caught up in criticizing what it does not like in the existing work of the economics profession and has not yet adequately focused on clearly defining an alternative. Any new theoretical approach grows out of accumulated dissatisfaction with existing work. And it is perhaps a truism that it is always easier to criticize what you do not like than to construct it in a different way. But at some point, if a perspective is to gain long-term legitimacy, it must shift from criticism to construction.[19]

While there is a place for pointing out that the *Titanic* is sinking even if one cannot be sure of finding lifeboats that work, there is also much to be said for Blank's counsel. This book is concerned with analysing the limits of existing work precisely in order to clarify an alternative. In Blank's terms the approach may concentrate too much on criticism and not enough on construction, and I suspect the outlined construction is not at all what she had in mind. Nevertheless, the argument presented here is a definite, if as yet sketchy, alternative. It proposes a chart with a rather different beginning, viewpoint and direction, and suggests both different ways of perceiving/employing conventional analytical techniques as well as the elaboration of new techniques. Because the argument suggests the navigation of comparatively uncharted seas, it does not, and cannot, predict what might issue from it, what new stories/'economyths' might be written in relation to a sexual epistemology of economics.

## Elaborating 'economyths' with women as the stories' subjects

The construction of alternative economyths, which move beyond the characteristic tendency of 'sexual economyths' to marginalise/exclude women and their specific experiences, is not likely to occur in isolation from other strands of feminist scholarship. That process of construction could make use of the work of feminist legal scholars in relation to their discussions of forms of legal recognition of the value of women's work and has some links with overlapping work by writers like Carol Gilligan who are concerned with the ways in which justice, morality and ethics may be reconceived in the light of women's involvement in 'care'.[20] There are further possible connections between an alternative feminist economics and developing approaches in political theory. Such approaches have suggested a reconsideration of the significance of the private sphere. In this context Carole Pateman notes that classical political philosophy describes the sphere of 'politics' in terms of a 'social contract' which frames the public arena and relatedly naturalises the operations of the private sphere as 'outside' politics. Pateman proposes a more inclusive understanding of politics which perceives the social contract associated with the organisation of the public realm as sustained by a 'sexual contract'.[21] That sexual contract in modern societies involves the fraternal exclusion of women from 'the social'. According to Yeatman, in a connected discussion, the.exclusion of women is accompanied by their positioning in relation to a distinctive 'expressive orientation' which is not identical with the attributes of the public arena.[22] In this analysis the classical liberal interpretation of politics as public may be considered to rest in a fundamental sense upon the 'invisible' dynamic of the private realm. Both Yeatman and Pateman delineate the dependence of civil society and the state on 'domestic despotism';[23] the 'sexual contract' is in these analyses a repressed dimension underlying the character of the modern public domain, integral to the very existence of that domain.[24]

It is possible that the alternative perspective in the field of economics described in this book could draw upon such accounts. These accounts imply by extension that definitions of economics, which have revolved around the market/public sphere, may not simply be expanded by recognition of a further distinct field—the specificity of the private sphere. The inclusion of that field might also alter the meaning of economics such that rather than the household economy always being seen as dependent on the market, the former may be viewed as a repressed dimension underlying the market economy and integral to its existence. Such possibilities certainly extend the impact of the approach taken in this book insofar as the implications of a conception of market economics being dependent on the system of the household economy in modern societies must significantly affect all existing economic paradigms. Even if one were to explore the potential of somewhat less dramatic epistemological reversals with regard to notions of dependence and propose more contextual, contingent forms of connection/dependence between the household and market economies,[25] this would still suggest that the delineation of the specificities of women's labour is likely to have consequences not only for a feminist economics but for assumptions regarding the entire field of 'economics' as it is usually understood. The point is that a useful extending feature of a perspective which perceives the limits of market economics and focuses upon the particularities of women's labour/sex relations and the household economy is that this perspective promotes examination of the importation of related feminist approaches attending to the dynamic of the private sphere.

The employment of feminist scholarship from fields other than economics (such as law, ethics, psychology and political theory) for the elaboration of a feminist economics may well speed the development of its fledgling parameters. Observations from other realms of feminist work dealing with the private sphere are likely to prove of particular assistance. Keller and Flax, for instance, link the evasion of the personal/private realm to a masculinist standpoint common to philosophies developed under social relations of male domination and associated with the masculine model of 'self' that maintains 'separation from and control over "others" in order to retain its own identity'.[26] Flax proposes that the masculine infantile experience of the sexual division of labour and society 'has had a deep and largely unexplored impact on philosophy. This repressed material shapes by its very absence in consciousness the way we look at and reflect upon the world'.[27] Flax is not suggesting that social theories are to be explained by reference to the particular psychology of individual philosophers nor that such themes are simply rationalisations of the unconscious.[28] She is, however, drawing attention to the intimate connection between public and private distinctions in modern (and possibly earlier) societies and theoretical constructions of social relations. Denial or underestimation of the private combined with an inclination to focus upon the public may reflect the dilemmas of hierarchical sex relations projected onto social life and therefore prove inadequate for a feminist perspective. Such accounts fill out the epistemological framework of an alternative perspective in feminist economics and strengthen the need to explore the specificity of the private realm.

The elaboration of that alternative is of course not restricted to the insights of contemporary feminist scholars. A perspective which centres on the particularities of the modern household economy/private sphere can also reassess other traditions of thought such as the work of early socialist/socialist feminist writers like Owen, Thompson, Wheeler and Bebel who 'emphasized the productivity of household labour and explored means of making it more efficient'.[29] Similarly, theorists who have been relatively excluded from discussions of literature dealing with the public/private divide might be subjected to closer scrutiny. Montesquieu, Mme de Staël and Jane Adams, for example, who offered 'a more subtle or positive view of women's domestic role' than widely employed writers like Hegel 'and/or proclaimed the relevance of domestic values to public life', may be seen as contributing to the ongoing development of an/other perspective on women's labours.[30]

Moreover, there are political or strategic areas of knowledge that might be brought to bear upon an alternative feminist approach to economics, and which can in turn draw upon that alternative. Knowledges which recognise forms of power in the private realm and discuss their capacity for translation or use in relation to the public sphere may also be relevant. In a world where it is increasingly conceivable that the public domain of the marketplace may be organised by a few vast organisations which reach beyond individual nation states, the social relations of the private sphere might be of some significance. Indeed their very *difference* from public forms of power may be politically relevant. The possibility of the private sphere acting as a site of instability, recalcitrance, contradiction or even resistance is worth consideration. Should this be viewed as an overly positive and/or dynamic account of the opportunities associated with private forms of power, I would note the example of 'the crucial role of Black family groupings in providing protection against the surrounding racism of white–dominated societies'.[30] There may be substantial obstacles to transferring private authority across the private/public divide but, as this instance indicates, it can at least be employed to contain the encroachments of public forms of power.

In this sketch of the possibilities for elaborating an alternative within feminist economics, which involves briefly drawing out resemblances with other approaches that focus on the private arena, there is no intention to celebrate in any uncritical sense the character of that domain. Relatedly, the critical view of market-based economics expressed here does not imply some construction of the household

as a haven for entirely desirable altruistic practices and values. This
book, as it moves towards an epistemological agenda for an alternative economic perspective, highlights not only the specificity of the
household economy but its specific mechanisms of expropriation. As
the poignant theme of the myth of Persephone indicates, the labours
of femininity are conjugated by both satisfactions and sorrows within
the power relations of the sexes. The crucial issue is not that the
'altruism'/emotionality of the household economy may suggest alternative, more edifying values than those of the market—though in
certain respects it may—but rather that it is an alternative economy
which requires analysis in its own terms. Though economic frameworks commonly employed by feminists, and many existing feminist
approaches, are oriented towards market-based or -derived analysis,
the elaboration of an alternative perspective is unlikely to arise in
frameworks indifferent to feminist concerns. The housework is yet
to be done and, for better or worse, it remains a labour in feminist
hands.

# Notes

## Introduction

1 G. Grass, *The Flounder: A Celebration of Life, Food and Sex*, Penguin, Harmondsworth, 1977, pp. 149–150.
2 M. Barrett, *Women's Oppression Today: Problems in Marxist Feminist Analysis*, Verso/NLB, London, 1980; M. Barrett and M. McIntosh, 'Christine Delphy: Towards a materialist feminism?', *Feminist Review*, no. 1, 1979, pp. 95–106; C. Delphy, 'A materialist feminism is possible', *Feminist Review*, no. 4, 1980, pp. 79–105.
3 S. Rees et al., 'Introduction', in *Beyond the Market: Alternatives to Economic Rationalism*, eds. S. Rees et al., Pluto Press, Leichhardt, NSW, 1993, pp. 7–10.
4 A. Touraine, 'Endgame', *Thesis Eleven*, no. 23, 1989, p. 124.
5 Cited in L. Bassi, 'Confessions of a feminist economist: Why I haven't yet taught an economics course on women's issues', *Women's Studies Quarterly*, vol. XVIII, nos. 3 & 4, 1990, p. 45, emphasis added.
6 Bassi, 'Confessions of a feminist economist', p. 43; P. Hyman, 'The use of economic orthodoxy to justify inequality: A feminist critique', in *Feminist Voices: Women's Studies Texts for Aotearoa/New Zealand*, ed. R. Du Plessis, Oxford University Press, Auckland, 1992, pp. 252–4; B. Bergmann, 'Feminism and economics', *Women's Studies Quarterly*, vol. XVIII, nos. 3 & 4, 1990, p. 69; R. Sharp and R. Broomhill, *Short-Changed: Women and Economic Policies*, Allen & Unwin, Sydney, 1988, p. 33; M. Power et al., 'Writing women out of the economy', paper prepared for the ANZAAS Centenary Congress, Sydney, May 17, 1986; M. Ferber and J. Nelson, 'Introduction: The social construction of economics and the social construction of gender', in *Beyond Economic Man: Feminist Theory and Economics*, eds. M. Ferber and J. Nelson, University of Chicago Press, Chicago, 1993, pp. 2–7.

7  Rees et al., 'Introduction', p. 10.
8  On women's waged work, see K. Mumford, *Women Working: Economics and Reality*, Allen & Unwin, Sydney, 1989; F. Blau and M. Ferber, *The Economics of Women, Men, and Work*, Prentice-Hall, Englewood Cliffs, N.J., 1986; B. Bergmann, *The Economic Emergence of Women*, Basic Books, N.Y., 1986; M. Ferber, 'Women and work: Issues of the 1980s', *Signs*, vol. 8, no. 2, Winter 1982, pp. 273–95.
   On links between market and subsistence work, see M. Mies, *Patriarchy and Accumulation on a World Scale: Women in the International Division of Labour*, Zed Books, London/Atlantic Heights, N.J., 1986; H. Afshar ed., *Women, Work and Ideology in the Third World*, Tavistock, London/N.Y., 1985; L. Beneria ed., *Women and Development: The Sexual Division of Labour in Rural Societies*, Praeger Pub., N.Y., 1985; K. Young et al. eds., *Of Marriage and the Market: Women's Subordination Internationally and Its Lessons*, RKP, London, 1984; E. Boserup, *Women's Role in Economic Development*, Allen & Unwin, London, 1970.
9  On household labour see, for example, M. Bittman, *Juggling Time: How Australian Families Use Time*, Office of the Status of Women, Department of the Prime Minister and Cabinet, Canberra, 1991; L. Morris, *The Workings of the Household*, Polity Press, Cambridge, 1990; Australian Bureau of Statistics, 'Measuring unpaid household work: Issues and experimental estimates', Information Paper, Catalogue No. 5236.0, Commonwealth Government, Canberra, 1990; D. Ironmonger ed., *Households Work: Productive Activities, Women and Income in the Household Economy*, Allen & Unwin, Sydney, 1989; H. Hartmann, 'The family as the locus of gender, class and political struggle: The example of housework', in *Feminism and Methodology: Social Science Issues*, ed. S. Harding, Indiana University Press/Open University Press, Bloomington, Indiana/Milton Keynes, 1987, pp. 109–37; A. Oakley, *The Sociology of Housework*, Pantheon, N.Y., 1974.
   On the relationship of household to waged work, see J. Baxter et al., *Double Take: The Links between Paid and Unpaid Work*, AGPS, Canberra, 1990; P. England and G. Farkas, *Households, Employment, and Gender: A Social, Economic and Demographic View*, Aldine, N.Y., 1986.
10  Sharp and Broomhill, *Short-Changed*, p. 37. Barbara Bergmann's 1990 compilation of reading lists from eight American economics courses which focus on women reinforces this point regarding the far greater attention given to market labour and public policy. B. Bergmann comp., 'Reading lists on women's studies in economics', *Women's Studies Quarterly*, vol. XVIII, nos. 3 & 4, 1990, pp. 75–86.
11  Jackson asserts that there are significant gaps in the literature on housework and that much of it is descriptive rather than analytical. S. Jackson, 'Towards a historical sociology of housework: A materialist feminist analysis', *Women's Studies International Forum*, vol. 15, no. 2, pp. 153–72.
12  Examples of conservative liberal (neoclassical) economic approaches include M. L. Greenhut, *Theory of the Firm in Economic Space*,

Meredith Corporation, N.Y., 1970; R. G. Lipsey, *An Introduction to Positive Economics*, Weidenfield & Nicolson, London, 1963; P. A. Samuelson, *Foundations of Economic Analysis*, Harvard University Press, Cambridge, 1948. Neoliberal economics itself consists of several strands, discussion of which is provided in D. Smith, *The Rise and Fall of Monetarism: The Theory and Politics of an Economic Experiment*, Harmondsworth, Penguin, 1987; W. Grant and S. Nath, *The Politics of Economic Policymaking*, Basil Blackwell, Oxford, 1984; M. Sawer ed., *Australia and the New Right*, Allen & Unwin, Sydney, 1982.

Examples of social liberal/social democratic approaches are H. Stretton, *Political Essays*, Georgian House, Melbourne, 1987; P. Wilenski, *Public Power and Public Administration*, Hale & Iremonger, Sydney, 1986; M. A. Jones, *The Australian Welfare State*, Allen & Unwin, Sydney, 1980. In referring to social liberals and social democrats under one heading I am describing a 'centrist' position which is distinguished by some subscription to humanist values but which can be delineated from Marxian views by its acceptance of a sociopolitical agenda that does not involve a radical restructuring of capitalism. (L. Bryson, *Welfare and the State*, Macmillan, London, 1992, pp. 40–1.) Some readers may be discomfited by the description of social liberal/social democratic economic approaches as 'mainstream', along with neoclassical economics, and consider with some justice that the latter is the prevailing paradigm in Western countries. I do not disagree but simply note that there is a well-established tradition of social liberal/social democratic perspectives in Western societies, evident for instance in the presence of political parties which broadly follow these perspectives. This is not the case for feminist accounts of economics. Even in Australia, where, according to Marion Sawer, feminists have been uniquely successful in entering and influencing the policy apparatus of the state, feminism remains a comparatively marginal approach in relation to overall economic thinking. M. Sawer, 'Why has the women's movement had more influence on government in Australia than elsewhere?', in *Australia Compared: People, Policies and Politics*, ed. F. Castles, Allen & Unwin, Sydney, 1991, pp. 258–77; M. Simms and D. Stone, 'Women's policy', in *Hawke and Australian Public Policy: Consensus and Restructuring*, eds. C. Jennett and R. Stewart, Macmillan, South Melbourne, 1990, pp. 294–7.

Examples of Marxian economic analyses include E. L. Wheelwright, 'Are the rich getting richer and the poor poorer? If so, why?', in *Questions for the Nineties*, ed. A. Gollan, Left Book Club, Sydney, 1990, pp. 199–215; C. Offe, *Disorganized Capitalism: Contemporary Transformations of Work and Politics*, ed. J. Kean, Polity Press, Cambridge, 1985; J. O'Connor, *The Fiscal Crisis of the State*, St. Martins Press, N.Y., 1973.

13 This point will be outlined in more detail later. See also Jennings for one discussion of the ways in which the market is privileged in conventional understandings of the distinction between the 'public' and 'private', a distinction critical to economic analyses. A. Jennings, 'Public or private? Institutional economics and feminism', in *Beyond Economic*

*Man*, eds. M. Ferber and J. Nelson, University of Chicago Press, Chicago, 1993, pp. 120–1.

14 Ferber and Nelson assert, for example, that most feminist *economists* consider that the problem of women's exclusion from/marginality within economics does not lie in prevailing (neoclassical) economic theory, but rather in the interpretation and use of that paradigm. The 'adjustment' to economic modes of analysis envisaged by feminist *economists* at least seems limited indeed. However, even the great majority of feminist scholars concerned with economics and women's labour, who are not economists and are critical of the neoclassical economic model, are inclined to remain largely within existing ('mainstream' or Marxian) frameworks, as will be shown in later chapters. Ferber and Nelson, 'Introduction: The social construction of economics and the social construction of gender', p. 8.

15 S. Lewenhak, *The Revaluation of Women's Work*, second edition, Earthscan Pub., London, 1992, pp. 8–10; M. Waring, *Counting for Nothing: What Men Value and What Women are Worth*, Allen & Unwin/Port Nicholson Press, Sydney, 1988, pp. 1–7.

16 N. Hartsock, *Money, Sex and Power*, Northeastern University Press, Boston, 1983; C. Delphy, *Close To Home*, Hutchinson, London, 1984.

17 See, for example, S. Gunew ed., *Feminist Knowledge: Critique and Construct*, Routledge, London/N.Y., 1990; C. Pateman, *The Sexual Contract*, Polity Press, Cambridge, 1988; C. Pateman and E. Gross eds., *Feminist Challenges: Social and Political Theory*, Allen & Unwin, Sydney, 1986; S. Harding and M. Hintikka eds., *Discovering Reality: Feminist Perspectives on Epistemology, Metaphysics, Methodology, and Philosophy of Science*, D. Reidel Pub., Dordrecht, 1983.

18 Ferber and Nelson note the comparative lack of feminist epistemological critiques of economics: '[m]ore recently, feminist theories have brought forth a substantial literature on questions of feminism and the pursuit of knowledge, but they have focused largely on the physical sciences and social sciences other than economics'. M. Ferber and J. Nelson, 'Preface', in *Beyond Economic Man*, eds. M. Ferber and J. Nelson, University of Chicago Press, Chicago, 1993.

19 This is Hyman's summary of Stilwell's view. Hyman, 'The use of economic orthodoxy to justify inequality', p. 254; F. Stilwell, 'Contemporary political economy: Common and contested terrain', *Economic Record*, vol. 64, no. 184, March 1988, pp. 14–25. Sharp and Broomhill also note that 'the impact of feminist critiques on mainstream economics has been relatively minor'. Sharp and Broomhill, *Short-Changed*, p. 39. In addition, see M. Ferber and M. Teiman, 'The oldest, the most established, the most quantitative of the social sciences—and the most dominated by men: The impact of feminism on economics', in *Men's Studies Modified: The Impact of Feminism on the Academic Disciplines*, ed. D. Spender, Pergamon Press, N.Y., 1981, pp. 125–39.

20 E. Cox, 'The economics of mutual support: A feminist approach', in *Beyond the Market*, ed. S. Rees et al., Pluto Press, Leichhardt, NSW, 1993, pp. 270–5.

21 The alternative perspective I propose probably differs in some respects

from Cox's approach in that she calls for 'social theories which assume that both private and public spheres are important', but her method for accomplishing this is through 'a public/private merger'. As will be outlined later, while I support Cox's general first point, I am less convinced that the best analytical or policy procedure for a feminist economics involves merging the public and private spheres. Cox, 'The economics of mutual support', pp. 275, 274.

22  *The Economics of Women, Men and Work* by Blau and Ferber is an exception.
23  Prue Hyman's 'The use of economic orthodoxy to justify inequality' is a rare example of a feminist critique of neoclassical economics. See also the first four essays in Ferber and Nelson, *Beyond Economic Man*, by Nelson, England, Strassmann and McCloskey, pp. 23–93.
24  Bassi, 'Confessions of a Feminist Economist', p. 42.
25  Bergmann, 'Feminism and Economics', p. 69.
26  See, for example, the integration of neoclassical perspectives with a range of other approaches in England and Farkas, *Households, Employment and Gender*.
27  However, since some household work studies could be said to employ neoclassical parameters, there is some discussion of issues that might involve direct application to orthodox economics. Additionally I deal briefly with certain neoclassical precepts in the conclusion.
28  C. Beasley, 'Can the contents of a 'tool box' do the housework?: considering the uses of a Foucauldian framework for investigating power and women's labour', paper presented at the Australian Sociological Association Conference, Adelaide, December 1992; C. Beasley, *The Sexual Metaphysics of Economics: A Feminist Critique of Post-modernism, Post-Marxism, Marxism and 'Materialist' Feminisms*, PhD thesis, Women's Studies, Flinders University of South Australia, 1991, (chapters 5, 6, 7 and 8 deal with post-Marxism).
29  Terry Eagleton, noting one plane of overlapping concerns, goes so far as to note the ongoing constitution of a 'certain historic alliance between feminism and post-structuralism, as radical oppositional movements gravely sceptical of central features of classical Marxist politics'. However, the wide range of feminist responses to and reworkings of the various strands of postmodernism goes far beyond this possible meeting point. T. Eagleton, 'Marxism, structuralism, and poststructuralism', *Diacritics*, vol. 15, no. 4, Winter 1985, p. 7; see also L. Nicholson ed., *Feminism/Postmodernism*, Routledge, N.Y., 1990; I. Diamond and L. Quinby eds., *Feminism and Foucault: Reflections on Resistance*, Northeastern University Press, Boston, 1988; J. Flax, 'Postmodernism and gender relations in feminist theory', *Signs*, vol. 12, no. 4, Summer 1987, pp. 621–43.
30  V. Beechey, *Unequal Work*, Verso, London, 1987, p. 13.
31  Rosemary Pringle is one example of an Australian writer who deals with women's labour and makes use of 'deconstructionist' approaches. R. Pringle, *Secretaries Talk: Sexuality, Power and Work*, Allen & Unwin, Sydney, 1988, pp. ix–xi.
32  Psychoanalytic feminist writers generally do not deal with women's

labour or economics explicitly, nor do feminist scholars dealing with economics and work typically engage with psychoanalytic theory (as Pringle argues). Ann Ferguson's work is one exception to this mutual lack of interaction. Ferguson draws upon psychoanalytic approaches, in particular writers like Luce Irigaray, and broadly employs these approaches amongst others in her analysis of a 'sex/affective production system'. A. Ferguson, *Blood at the Root: Motherhood, Sexuality and Male Dominance*, Pandora Press, London, 1989, Chapters 1–4; Pringle, *Secretaries Talk*, p. ix.

33  V. S. Peterson ed., *Gendered States: Feminist (Re)Visions of International Relations Theory*, Lynne Rienner, Boulder, Colorado, 1992; R. Grant and K. Newland eds., *Gender and International Relations*, Millennium Pub., London, 1991; see also Sharp and Broomhill, *Short-Changed*, p. 38 and this chapter, note 8.

34  M. Morokvasic, 'Fortress Europe and migrant women', *Feminist Review*, no. 39, 1991, pp. 69–84; C. Enloe, 'Silicon tricks and the two dollar woman', *New Internationalist*, January 1992, pp. 12–14; Mies, *Patriarchy and Accumulation on a World Scale*.

35  J. Pettman, 'Women, nationalism and the state: Towards an international feminist perspective', paper presented for the Australian Women's Studies Association Conference, University of Sydney, 1992, p. 12; see also T. Minh-ha, 'Difference—a special Third World issue', *Feminist Review*, no. 25, 1987, pp. 5–20.

36  F. Mascia-Lees et al., 'The postmodernist turn in anthropology: Cautions from a feminist perspective', *Signs*, vol. 15, no. 1, Autumn 1989, p. 8. I follow Salleh's account of terms like women (woman), feminine and female here. (K. Salleh, 'Contribution to the critique of political epistemology', *Thesis Eleven*, no. 8, January 1984, pp. 40–1.)

> [T]he terms *masculine* and *feminine* are used . . . where *historically produced* or culturally defined gender types are implied, while the terms *female* and *male* are used where reference is made to *biologically located* sex differences . . . [The intention], however, is to break down all such dualisms as hard and fast categories by emphasizing their reciprocal conditioning. The designations, *woman* and *man*, are commonly used to describe a blend of nature and culture, biology and history, but confusingly, these must carry not only an *immanent* or everyday ideological sense, but a *transcendent* or potentially critical one as well.

37  J. Flax, 'Postmodernism and gender relations in feminist theory', *Signs*, vol. 12, no. 4, Summer 1987, pp. 622–3.

38  Morris insists that feminism has no obligations to postmodernism. Pringle makes the same comment with reference to Marxism. M. Morris, *The Pirate's Fiancée: Feminism, Reading and Postmodernism*, Verso, London/N.Y., 1988, p. 12; R. Pringle,' "Socialist-Feminism' in the eighties: Reply to Curthoys', *Australian Feminist Studies*, no. 6, Autumn 1988, p. 27.

39  A. Edwards, 'The sex/gender distinction: Has it outlived its usefulness?', *Australian Feminist Studies*, no. 10, Summer 1989, pp. 1–12; D. Thompson, 'The sex/gender distinction: A reconsideration', *Australian Feminist Studies*, no. 10, Summer 1989, pp. 23–31.

40   Edwards, 'The sex/gender distinction, p. 8; J. Butler, *Gender Trouble: Feminism and the Subversion of Identity*, Routledge, N.Y./London, 1990, pp. 6–25.
41   N. Hartsock, 'The feminist standpoint: Developing the ground for a specifically feminist historical materialism', in *Feminism and Methodology: Social Science Issues*, ed. S. Harding, Indiana University Press/Open University Press, Bloomington, Indiana/Milton Keynes, 1987, p. 163.
42   J. Allen and E. Grosz, 'Editorial', *Australian Feminist Studies* ('Feminism and the body'), no. 5, Summer 1987, p. viii; E. Grosz, 'Notes towards a corporeal feminism', *Australian Feminist Studies*, no. 5, Summer 1987, p. 4; C. Pateman and M. Shanley, 'Introduction', in *Feminist Interpretations and Political Theory*, eds. M. Shanley and C. Pateman, Polity Press, Cambridge, 1991, p. 3.
43   Nevertheless the term 'gender' may have other advantages. In particular it may be used to disrupt assumptions about a causal nexus between bodies, gender identity and desire. See in this context Butler's usage of 'gender'. Butler, *Gender Trouble*; J. Butler, 'Contingent foundations: Feminism and the question of postmodernism" ', in *Feminists Theorize the Political*, eds. J. Butler and J. Scott, Routledge, N.Y./London, 1992, pp. 3–21.
44   Butler, *Gender Trouble*, p. 30.
45   E. Gross, 'Philosophy, subjectivity and the body: Kristeva and Irigaray', in *Feminist Challenges: Social and Political Theory*, eds. C. Pateman and E. Gross, Allen & Unwin, Sydney, 1986, p. 136.
46   Gross and Gatens both draw attention to the political significance of different kinds of bodies. In this context I find helpful their discussion of femininity and its relation to male and female bodies. However it is possible that they assume an overly clear-cut and self-evident distinction between male and female bodies and do not allow sufficiently, for example, various forms of hermaphroditism. For this reason it seems to me that notions of 'kinds' of bodies must always be less than absolute. Gross, 'Philosophy, subjectivity and the body', p. 13; M. Gatens, 'A critique of the sex/gender distinction', in *Beyond Marxism: Interventions after Marx*, eds. J. Allen and P. Patton, Intervention Pub., Leichhardt, NSW, 1983, pp. 148–9 and 153–4.
47   Gatens, 'A critique of the sex/gender distinction', p. 148.
48   Grass, *The Flounder*, p. 168.

## 1 The Marx Question

1   J. Maynard Keynes, quoted in J. M. Cohen and M. J. Cohen, *The Penguin Dictionary of Modern Quotations*, second edition, Penguin, Harmondsworth, 1988, p. 183.
2   S. Conran, quoted in Cohen and Cohen, *The Penguin Dictionary of Modern Quotations*, p. 85.
3   T. Eagleton, 'Base and superstructure in Raymond Williams', in T. Eagleton ed., *Raymond Williams: Critical Perspectives*, Polity Press, 1989, p. 169.
4   S. Hall, 'The 'political' and the 'economic' in Marx's theory of classes',

in *Class and Class Structure*, ed. A. Hunt, Lawrence and Wishart, London, 1977, pp. 22–3; R. Lichtman, 'Marx's theory of ideology', *Socialist Revolution*, vol. 5, no. 23, April 1975, pp. 46 and 51.

5  C. Johnson, 'The transformation of proletarian consciousness in Marx's theory of revolution', unpublished M.A. Econ. thesis, University of Manchester, 1978, passim. Though, as Lichtman has noted, 'the analysis of fetishism and ideology presented in *Capital* undermines any easy enthusiasm', Marx continues to argue that with the advance of productive forces there will develop, in a relatively straightforward fashion, a revolutionary consciousness. This is particularly evident in the chapter on 'Commodities' in Volume I of *Capital*, where he quotes without alteration a section from *The Poverty of Philosophy*. Marx's views of the 1860s and 1870s appear to remain remarkably consistent with those of the 1840s to 50s. As Fernbach has noted in his discussion of Marx's letter to Bolte of November 1871:

> It is surprising how firmly Marx maintains in this letter the position on the development of the proletarian movement put forward in the Manifesto, despite the experience of the intervening years . . . No more than in the Manifesto does Marx leave theoretical space for the possibility of a workers' movement that is organized politically as a class and yet struggles solely for reforms within the capitalist system.

Lichtman, 'Marx's theory of ideology', p. 71; D. Fernbach, 'Introduction', in K. Marx, *The First International and After*, ed. D. Fernbach, vol. 3, Penguin/NLB, Harmondsworth, 1974, p. 59

6  E. Laclau and C. Mouffe, 'Socialist strategy: Where next?', *Marxism Today*, vol. 25, no. 1, January 1981, p. 17; see also E. Laclau and C. Mouffe, 'Post-Marxism without apologies', *New Left Review*, no. 166, November/December 1987, pp. 86–90.

7  Eagleton, 'Base and superstructure', pp. 172–3; see also P. Beilharz, 'Marxism and history', *Thesis Eleven*, no. 2, 1981, p. 16.

8  This can be said to occur, for example, in readings of *The German Ideology*. Writers like Lichtman and Hall presume that this work propounds a general theory of ideology. Lichtman argues that Marx here produces a simple copy theory of knowledge derived from a simplistic construction of base/superstructure and demonstrates this by reference to the 'camera obscura' metaphor. While there may be some foundation to the view that Marx in *The German Ideology* adopts a rather unsophisticated version of the base/superstructure model, it is clear from the paragraph which follows the metaphor that Marx is not offering a theory of the superstructure as a reflection of the 'base' but referring to his general ontological doctrine concerning consciousness and social being, and in this context noting his divergence from the flawed particularities of German speculative philosophy: '[i]n direct contrast to German philosophy which descends from heaven to earth here it is a matter of ascending from earth to heaven'. Lichtman, 'Marx's theory of ideology', p. 46; K. Marx and F. Engels, *The German Ideology*, Progress Pub., Moscow, 1976, p. 42.

9  Eagleton, 'Base and superstructure', p. 172.

10  ibid., p. 173.

11 See this chapter, note 22, for discussion of questions associated with Eagleton's method of dealing with the slippage between ontological and historical frameworks.

12 L. Nicholson, 'Feminism and Marx: Integrating kinship with the economic', in *Feminism as Critique: Essays on the Politics of Gender in Late Capitalist Societies*, eds. S. Benhabib and D. Cornell, Polity Press, Cambridge, 1987, pp. 19–20.

13 Marx does not necessarily conflate his perception of determinacy with a perspective that the economic is entirely distinct from the 'superstructural', for even in the early works such as *The German Ideology* he explicitly states that consciousness is bound up with the process of production. Nor does it appear that Marx refuses in a clear-cut fashion to grant *any* efficacy to the 'superstructure'. As is obvious from Marx's reply to a critic of the 1859 Preface to *A Contribution to the Critique of Political Economy*, he does allow that in certain historical periods the 'super-structure' could come to play a *dominant* role within societies; for example in classical antiquity and in feudal society, politics and religion respectively are claimed to have played 'the chief part' in the development of the specific character of these periods . Moreover, it can be argued that Marx does give the 'superstructure' an 'objective status', even if in the case of revolutionary consciousness that status is depicted as dissolving before the developing conditions of capitalist economics. Marx and Engels, *The German Ideology*, p. 95; S. Hall, 'Rethinking the 'base-and-superstructure' Metaphor', in *Class, Hegemony and Party*, eds. J. Bloomfield et al., Communist University of London, Lawrence and Wishart, London, 1976, p. 49; K. Marx, *Capital*, vol. 1, Penguin, Harmondsworth, 1976, pp. 175–76; J. Rancière, 'On the theory of ideology (the politics of Althusser)', *Radical Philosophy*, no. 7, Spring 1974, p. 9; see also this chapter, note 5.

14 C. Fritzell, 'On the concept of relative autonomy in educational theory', *British Journal of Sociology of Education*, vol. 8, no. 1, 1987, p. 26.

15 D. Ironmonger, 'Preface', in *Households Work: Productive Activities, Women and Income in the Household Economy*, ed. D. Ironmonger, Allen & Unwin, Sydney, 1989, pp. ix–x.

16 C. Johnson, Review (women and politics), *Australian Feminist Studies*, no. 1, Summer 1985, p. 144.

17 An example of Laclau and Mouffe's inclination to perceive the economic/production/labour in ways that illustrate an acceptance of Marxian meanings of these terms is relevant here. While they assert the ways in which their approach challenges Marxism, they often display assumptions which reveal a congruence with that tradition. This is evident in the following example, where they move from plurality to class-based conceptions of production and power: '[f]ar from forming a homogeneous field ruled by the simple logic of profit maximisation, the economy is in actual fact a complex relation of forces between various social agents, and the productive forces are themselves subject to the rationality imposed on them by the ruling class'. (Laclau and Mouffe, 'Socialist strategy', p. 22) 'The economy' is certainly understood in a

more complex and 'culturalist' manner than in many Marxist accounts but, though it is no longer homogeneous, its domain appears unchanged. (This tendency is also evident in Touraine and Offe's work. See A. Touraine, 'Is sociology still the study of society?', *Thesis Eleven*, no. 23, 1989, pp. 5–7; C. Offe, *Disorganized Capitalism: Contemporary Transformations of Work and Politics*, ed. J. Keane, Polity Press, Cambridge, 1985, pp. 63, 96–7.)

Laclau and Mouffe's tendency to remain within the bounds of Marxist understandings of economics is further revealed in their focus on politics. They imply that the only way that feminist demands can be adequately included in a progressive framework is by a concentration on politics. Politics can encompass women, whereas Marxism's concentration on economics cannot: 'power does not derive from a place in the relations of production'. 'Sexist' and 'patriarchal' forms of social organisation are not apparently covered by 'economics' and therefore notions of economic determinacy must be abandoned and the primacy of politics within economics itself is postulated. But this argument does not follow *unless* one equates 'economics' with the 'relations of production' as described by Marxism. It is certainly possible to include sex relations in a social analysis which proposes a notion of economic determinacy where economics is not tied to a class-bound model. Christine Delphy, for example, provides exactly such an economic theory, outlining a 'base' which is not identical with the 'production' relations of capitalism. (Laclau and Mouffe, 'Socialist strategy', pp. 21–2; C. Delphy, 'Continuities and discontinuities in marriage and divorce', in *Sexual Divisions and Society: Process and Change*, eds. D. Barker and S. Allen, Tavistock, London, 1976, pp. 76–89; C. Delphy, 'A materialist feminism is possible', *Feminist Review*, no. 4, 1980, pp. 79–105.)

18  Eagleton, 'Base and superstructure', p.166.
19  ibid.
20  ibid., p. 167.
21  ibid., p. 167.
22  There are some additional reasons for disquiet at the given status accorded 'exploititative economic production'. Eagleton's explanation for the logic of economic causality proposed by the Marxist base/superstructure model lies in a conception of history unified by the exploitation of labour. This explanation imposes an intentional flattened and static quality upon the vagaries of social relations. History thus far is conceived as a barren, featureless terrain ultimately devoid of fluidity, a moonscape of arid subordination. Even those radical feminists who might be attracted to an analytical framework based upon singular causality and the transhistorical identity of past social relations around notions of relentless male dominance might quake before the starkness of Eagleton's vision of Marxism's historical perspective. Additionally, many feminists, including orthodox Marxist feminists, acknowledge that traditional communal societies in particular may offer an image of sociality relatively free of subordination; that is they propose a historical analysis which does not easily fit with any blanket statement. Indeed classical Marxism itself appears to have provided a description of history that differs in

certain respects from the uniformity which Eagleton construes as critical to Marxism. Marx and Engels's depictions of pre-class societies suggest that history is not all of a piece and that subordination may vary over time. This, of course, does not contradict Eagleton's view of the unity of history, but indicates that in Marxism volatility is upheld *alongside* the determinacy of exploitation: the latter is not perhaps as completed as Eagleton's approach suggests. Consequently, Eagleton's defence of the base/superstructure model as founded in the unity of history does appear to be somewhat weakened. Moreover, if Eagleton's defence is accepted and Marx's concept of generalised economic causality is taken as given, there is a cost, of which Eagleton is well aware. In his attempt to convince readers of the persuasiveness of the Marxist paradigm, he must assert the uniformity of history. The inevitable result is the loss of a considerable degree of historical fluidity—even allowing for Marx's own attempts to insist upon a degree of flexibility/change. Post-Marxists, given their emphasis on the creative opportunities of politics, remain unconvinced that this price is worth paying. While I would not abandon the potential constraints of notions of generality or causality in history per se, I think they have a point.

(Neither Engels nor Marxist feminist approaches that follow his perspective provide accounts consistent with a perception of all history as unending and consistent toil in exploitative circumstances. F. Engels, *The Origin of the Family, Private Property and the State*, International Pub., N.Y., 1972, p. 113; E. Zaretsky, 'Socialism and feminism I: Capitalism, the family, and personal life, part I', *Socialist Revolution* 3, nos. 1/2, January-April 1973, pp. 69–125.)

23  In this context Folbre points out that the analysis of class exploitation developed by Marx and Engels was defined, largely in terms of the interests of working-class men.

> This theory 'incorporated basic elements of the outlook of the organized, mainly skilled working men of the 1840s, including the male worker's conception of himself as the sole, rightful breadwinner for the working class family' . . . Neither Marx nor Engels disagreed that women were oppressed, but they linked this oppression to the consequences of private property and the interests of capital rather than to men's interests or men's power. As a result, they believed that resolution of 'the woman issue' could be achieved only by resolution of 'the class issue'. Efforts to reverse, or even to equalize, these priorities were considered counterproductive . . . In a letter to a friend in the late 1860s, Marx spoke derisively of women's suffrage, explaining that 'German women should have begun by driving their men to self-emancipation' rather than 'seeking emancipation for themselves directly'.

Folbre, 'Socialism, feminist and scientific', in *Beyond Economic Man: Feminist Theory and Economics*, eds. M. Ferber and J. Nelson, University of Chicago Press, Chicago, 1993, pp. 102–3.
24  Eagleton, 'Base and superstructure', p. 173.
25  J. Mitchell, *Psychoanalysis and Feminism*, Penguin, Harmondsworth, 1975.
    J. Mitchell, 'Introduction I', in *Feminine Sexuality: Jacques Lacan and*

*the* École Freudienne, eds. J. Mitchell and J. Rose, trans. J. Rose, Macmillan, London, 1982, pp.1–26; P. Adams and E. Cowie, 'Feminine Sexuality: Interview with Juliet Mitchell and Jacqueline Rose', *m/f*, no. 8, 1983, pp. 3–16.

26  I. Young, 'Beyond the unhappy marriage: A critique of the dual systems theory', in *Women and Revolution: A Discussion of the Unhappy Marriage of Marxism and Feminism*, ed. L. Sargent, South End Press, Boston, 1981, pp. 46–7.

27  This critique refers to two works by Barrett. M. Barrett, *Women's Oppression Today: Problems in Marxist Feminist Analysis*, Verso/NLB, London, 1980; M. Barrett and M. McIntosh, 'Christine Delphy: Towards a materialist feminism?', *Feminist Review*, no. 1, 1979, pp. 95–106.

More detailed discussion of difficulties in Barrett's approach is provided in J. Allen, 'Marxism and the man question: Some implications of the patriarchy debate', in *Beyond Marxism?: Interventions after Marx*, eds. J. Allen and P. Patton, Intervention Pub., Sydney, 1983, pp. 91–111; C. Beasley, 'The patriarchy debate: should we make use of the term 'patriarchy' in historical analysis?', *History of Education Review*, vol. 16, no. 2, 1987, pp. 13–20; Delphy, 'A materialist feminism is possible', pp. 92–102.

28  R. Tong, *Feminist Thought: A Comprehensive Introduction*, Unwin Hyman, London, 1989, p. 61; J. Elshtain, *Public Man, Private Woman*, Princeton University Press, Princeton, N.J., 1981.

29  A. Jaggar, *Feminist Politics and Human Nature*, Rowman and Allanheld, Totowa, N.J., 1983, p. 221.

30  Tong, *Feminist Thought*, p. 64.

31  H. Hartmann, 'The unhappy marriage of marxism and feminism: Towards a more progressive union', in *Women and Revolution: A Discussion of the Unhappy Marriage of Marxism and Feminism*, ed. L. Sargent, South End Press, Boston, 1981, p. 6.

32  Allen, 'Marxism and the man question', passim; M. Campioni and E. Gross, 'Love's labours lost: Marxism and feminism', in *Beyond Marxism?: Interventions after Marx*, eds. J. Allen and P. Patton, Intervention Pub., Sydney, 1983, passim.

33  Barrett considers that sex relations are historically embedded in a necessary reproduction of class relations, and has been accused on this basis of privileging the latter (see Allen, 'Marxism and the man question'). Yet at the same time she acknowledges that sex relations are *analytically* distinct from class relations, that class cannot adequately account for the relation of the sexes and that aspects of sex are simply irreducible to the demands of surplus value. Consequently she is critical of Marxist feminist analyses which assume a causal and functional fit between sex and class relations, such as those by Kuhn and Wolpe, Eisenstein, and McDonagh and Harrison. Barrett, *Women's Oppression Today*, pp. 24–9.

34  Marx presents and politicises only some attributes of the subject/body. In this context the limits of his approach may be usefully compared with the different limits of other frameworks. Marx's analysis has specific restrictions, as do the accounts of other writers. The subject within

Foucault's analyses is, for example, curiously individualised. Little attention is given to intersubjective or collective attributes/capacities of the
self. This is in contrast to the concern with (at least certain kinds of)
collective and socially interactive qualities of the subject in Marx's works.
Moreover, as Grosz notes, embodied subjectivity in Foucauldian analysis
'can be seen as a surface, an externality that presents itself to others
and to culture as a *writing* or inscriptive surface'. She compares this
trajectory with that of Freud and Lacan, who explore the sexualised
body from the 'inside'. On this basis Foucault's approach implies some
difficulties in describing interactions between 'inner' and 'outer' aspects
of subjectivity. These interactions may be perceived, however, in Marx's
active account of the labouring self. (See on intersubjective and collective
subjectivity, N. Fraser, 'Towards a discourse ethic of solidarity', *Praxis
International*, vol. 5, no. 4, January 1986, pp. 427–8; S. Benhabib and
D. Cornell, 'Introduction: Beyond the politics of gender', in *Feminism
as Critique*, eds. S. Benhabib and D. Cornell, Polity Press, Cambridge,
1987, op. cit., pp. 10–11; M. Foucault, *Language, Counter-Memory,
Practice*, ed. D. Bouchard, Cornell University Press, N.Y., 1977, pp.
216–7; C. Poynton, 'The privileging of representation and the marginalizing of the interpersonal: a metaphor (and more) for contemporary gender
relations', in *Feminine Masculine and Representation*, eds. T. Threadgold
and A. Cranny-Francis, Allen & Unwin, Sydney, 1990, passim.) For
Grosz's view of Foucault, Freud and Lacan, see E. Grosz, 'Notes towards
a corporeal feminism', *Australian Feminist Studies*, no. 5, Summer 1987,
pp. 9–10.)

35  Marx's analysis of the fetishism of commodities denotes a subjectivity
that is not wholly conscious or 'rational'.

36  It is undoubtedly of concern here that many feminist writers who doubt
the usefulness of Marx's perspective with regard to women's labour and
social positioning assume that this perspective is quite adequate to
describe waged labour and class relations. As Tong points out in relation
to Hartmann's dual systems model, Hartmann's account of capitalism is
the standard one of orthodox Marxism. Thus the model fails to challenge
Marxism's hegemony over studies of economics and labour as thoroughly
as might be expected. However, my focus, like that of Hartmann, is on
the relations of the sexes and therefore the implications of a viewpoint
indicating the limits of Marxism qua Marxism are merely suggestive.
Tong, *Feminist Thought*, p. 180.

37  Delphy, 'Continuities and discontinuities in marriage and divorce', p. 78.

38  Hartmann, 'The unhappy marriage of marxism and feminism', pp.
10–11.

39  J. Goodnow, 'Work in households: an overview and three studies', in
*Households Work: Productive Activities, Women and Income in the
Household Economy*, ed. D. Ironmonger, Allen & Unwin, Sydney, 1989,
p. 39; D. Ironmonger, 'Households and the household economy', in
*Households Work, ed. D. Ironmonger, Allen & Unwin, Sydney, 1989,
p. 3.

40  A. Ferguson, 'On conceiving motherhood and sexuality: A feminist
materialist approach', in *Mothering: Essays in Feminist Theory*, ed. J.

Trebilcot, Rowman and Allanheld, Totowa, N.J., 1984, pp. 153–82; D. Smith, 'Women, class and family', *Socialist Register*, 1983, pp. 1–44; M. O'Brien, *The Politics of Reproduction*, RKP, London/Boston, 1981; N. Hartsock, *Money, Sex, and Power: Toward a Feminist Historical Materialism*, Longman, N.Y., 1983; C. MacKinnon, 'Feminism, Marxism, method, and the state: An agenda for theory', *Signs*, vol. 7, no. 3, Spring 1982, pp. 515–44; Young, 'Beyond the unhappy marriage', pp 43–69; Jaggar, *Feminist Politics and Human Nature*; Hartmann, 'The unhappy marriage'; C. Delphy, *Close to Home: A Materialist Analysis of Women's Oppression*, trans. and ed. D. Leonard, Hutchinson, London, 1984.

41  Ironmonger asserts that 'altruism' is the motivating force in the household economy but uses this term without any apparent equivocation or awareness of the mystificatory effects of describing women's work as altruistic. The language of love, affection, and altruism must be employed cautiously so as not to slide into acceptance of the voluntarist, naturalised connotations of these words or to evade the coercive aspects of their character. Ironmonger, 'Households and the household economy', p. 3.

42  Delphy, 'Continuities and discontinuities in marriage and divorce', pp. 76–89.

43  Quotation from M. Thompson, 'Comment on Rich's 'Compulsory heterosexuality and lesbian existence'', *Signs*, vol. 6, no. 4, Summer 1981, pp. 791–2; C. Beasley, *The Ambiguities of Desire: Patriarchal Subjectivity*, unpublished M.A. thesis, Centre for Contemporary Cultural Studies, University of Birmingham, U.K., 1985, p. 112.

44  M. Draper, 'Women in the home', in *Households Work*, ed. D. Ironmonger, Allen & Unwin, Sydney, 1989, pp. 87–9.

45  J. Allen, Review (Hartsock), *Signs*, vol. 10, no. 3, Spring 1985, p. 578; see also Hartsock, *Money, Sex, and Power*.

46  G. Rubin, 'The traffic in women: Notes on the 'political economy' of sex', in *Toward an Anthropology of Women*, ed. R. Reiter, Monthly Review Press, N.Y./London, 1975, pp. 157–210; Delphy, *Close to Home*.

47  MacKinnon, 'Feminism, Marxism, method and the state: An agenda for theory', p. 516.

48  MacKinnon's approach as discussed in J. Miller, 'Comments on MacKinnon's 'Feminism, Marxism, Method and the State'', *Signs*, vol. 10, no. 1, Autumn 1984, p. 170.

49  Miller, 'Comments on MacKinnon's 'Feminism, Marxism, method and the state'', pp. 170–1; J. Acker and K. Barry, 'Comments on MacKinnon's 'Feminism, Marxism, method and the state'', *Signs*, vol. 10, no. 1, Autumn 1984, p. 176; C. MacKinnon, 'Reply to Miller, Acker and Barry, Johnson, West, and Gardiner', *Signs*, vol. 10, no. 1, Autumn 1984, pp. 184–6.

50  Jaggar, *Feminist Politics and Human Nature*, p. 353; Tong, *Feminist Thought*, p. 186.

51  MacKinnon, 'Feminism, Marxism, method and the state: An Agenda for Theory', p. 542; Miller, 'Comments on MacKinnon's "Feminism, Marxism, method and the state"', p. 171.

52  M. Hintikka, and J. Hintikka, 'How can language be sexist?', in

*Discovering Reality: Feminist Perspectives on Epistemology, Metaphysics, Methodology, and Philosophy of Science*, eds. S. Harding and M. Hintikka, D. Reidel Pub., Dordrecht, 1983, p. 146.

53 J. Fowles, *The Magus*, quoted in Cohen and Cohen, *The Penguin Dictionary of Modern Quotations*, pp. 120–1.

54 K. Marx, *Capital*, vol. 1, pp. 660 and 431.

55 ibid., p. 659.

56 Goodnow, 'Work in households', p. 46.

57 ibid.

58 J. Jacquette, 'Power as ideology: A feminist analysis', in *Women's Views of the Political World of Men*, ed. J. Stiehm, Transnational Pub., N.Y., p. 26.

59 M. Edwards, 'Commentary', in *Households Work*, ed. D. Ironmonger, Allen & Unwin, Sydney, 1989, pp. 35–6.

60 G. Cohen, *Karl Marx's Theory of History: A Defence*, Princeton University Press, Princeton, 1978, p. 134; G. McLennan, 'Philosophy and history: some issues in recent marxist theory', in *Making Histories: Studies in history-writing and politics*, eds. R. Johnson et al., Hutchinson/CCCS, London/Birmingham, 1982, pp. 136–7.

61 Marx, *Capital*, vol., pp. 284 and 287.

62 This is a rephrasing of Gross and Averill's view of what they see as patriarchal themes of scarcity and competition in relation to nature and in evolutionary theory. M. Gross and M. Averill, 'Evolution and patriarchal myths of scarcity and competition', in *Discovering Reality*, eds. S. Harding and M. Hintikka, D. Reidel Pub., Dordrecht, 1983, pp. 71–95.

63 P. Smith, 'Domestic labour and Marx's theory of value', in *Feminism and Materialism: Women and Modes of Production*, eds. A. Kuhn and A. Wolpe, RKP, London, 1978, pp. 206–8.

64 Murphy's review of Turner's work draws attention to this feature of women's emotional work when he points out that certain aspects of nursing in waged labour are unable to be entirely routinised, bureaucratised, ordered or technicised. J. Murphy, Review (Turner), *Thesis Eleven*, no. 22, 1989, p. 129; B. Turner, *Medical Power and Social Knowledge*, Sage, London, 1987.

65 Smith, 'Domestic labour and Marx's theory of value', passim.

66 ibid., p. 213. Smith asserts this point in relation to a debate with Gough. Gough notes that Marx uses 'a historical perspective' to determine the labour necessary to produce a given use value, but paradoxically ignores contextual conditions when assessing the 'necessity' of the final use value. According to Gough, for Marx the productiveness of labour is dependent on the former methodology. Smith finds nothing problematic in this. Marx, he argues, is attempting to analyse how production is understood in capitalism and on this basis rejects Gough's critique: 'Gough ignores the fact that Marx's definition of productive labour in the capitalist mode of production is made from the standpoint of capital, not from the point of view of Gough's blueprint for socialist production.'

## 2 Getting more specific

1 C. Porter, 'My heart belongs to Daddy', song from the musical, 'Leave it to me', quoted in J. M. and M. J. Cohen, *The Penguin Dictionary of Modern Quotations*, second edition, Penguin, Harmondsworth, 1980, p. 267.

2 B. Disraeli, Speech in the British House of Commons, 28 February 1859, quoted in J.M. Cohen and M.J. Cohen, *The Penguin Dictionary of Quotations*, Penguin, Harmondsworth, 1960, p. 140.

3 C. Delphy, 'Continuities and discontinuities in marriage and divorce', in *Sexual Divisions and Society: Process and Change*, eds. D. Barker and S. Allen, Tavistock, London, 1976, p. 86.

4 N. Folbre, 'Socialism, feminist and scientific', in *Beyond Economic Man: Feminist Theory and Economics*, eds. M. Ferber and J. Nelson, University of Chicago Press, Chicago, 1993, p. 103; M. O'Brien, *The Politics of Reproduction*, RKP, London/Boston, 1981; L. Clark and L. Lange, *The Sexism of Social and Political Theory: Women and Reproduction from Plato to Nietzsche*, University of Toronto Press, Toronto, 1979.

5 P. Smith, 'Domestic labour and Marx's theory of value', in *Feminism and Materialism: Women and Modes of Production*, eds. A. Kuhn and A. Wolpe, RKP, London, 1978, pp. 207, 208 and 210–11.

6 C. Johnson, 'Some problems and developments in Marxist feminist theory', in *Working it Out: All her Labours*, ed. Women and Labour Publications Collective, Hale and Iremonger, Marrickville, NSW, 1984, p. 128.

7 K. Marx, *Capital*, vol. 1, Penguin, Harmondsworth, 1976, pp. 284 and 287.

8 In the 1970s Marxist and feminist theorists debated whether domestic labour was in Marxist terms 'value producing' labour or not and in the process raised certain problems with the applicability of value theory to analysis of women's upaid work. Johnson, 'Some problems and developments in Marxist feminist theory', pp. 127–8; M. Eichler, *The Double Standard: A Feminist Critique of Feminist Social Science*, Croom Helm, London, 1980, pp. 111–3; Smith, 'Domestic labour and Marx's theory of value'; Folbre, 'Socialism, feminist and scientific', p. 103.

9 Marx, *Capital*, vol. 1, pp. 284 and 287; see section entitled 'The fetishism of commodities and the secret thereof.'

10 F. Maas, 'Commentary', in *Households Work: Productive Activities, Women and Income in the Household Economy*, ed. D. Ironmonger, Allen & Unwin, Sydney, 1989, p. 15.

11 M. Thornton, 'Commentary', in *Households Work*, ed. D. Ironmonger, Allen & Unwin, Sydney, 1989, p. 59.

12 J. Goodnow, 'Work in households: an overview and three studies', in *Households Work*, ed. D. Ironmonger, Allen & Unwin, Sydney, 1989, p. 39. On the other hand, Goodnow does not dismiss Oakley's use, for example, of an aggregate approach which proposes adding up the various activities of housewives into a single sum. Thornton, 'Commentary', pp. 59–60; A. Oakley, *The Sociology of Housework*, Pantheon books, N.Y., 1974.

13  Goodnow, 'Work in households', p. 39.
14  ibid., p. 40.
15  Thornton, 'Commentary', p. 59; Oakley, *The Sociology of Housework*, passim.
16  Goodnow, 'Work in households', pp. 49–50.
17  D. Ironmonger, 'Households and the household economy' in *Households Work*, ed. D. Ironmonger, Allen & Unwin, Sydney, 1989, pp. 6–7
18  See for another example of the use of this distinction, Australian Bureau of Statistics (ABS), 'Measuring unpaid household work: Issues and experimental estimates', ABS Information Paper, Catalogue No. 5236.0, Commonwealth Government, Canberra, 1990; see also comparisons with international studies from Canada, USA, France and Finland ibid., p. 17.
19  D. Ironmonger and E. Sonius, 'Household productive activities', in *Households Work*, ed. D. Ironmonger, Allen & Unwin, Sydney, 1989, pp. 24–5.
20  ibid., p.25.
21  ibid., p.25.
22  Ironmonger, 'Households and the household economy', pp. 5–6; S. Burns, *The Household Economy: Its Shape, Origins and Future*, Beacon Press, Boston, 1977, p. 8.
23  In 1981, the ILO adopted Convention No. 156 and Recommendation No. 165 concerning Workers with Family Responsibilities. Convention 156 arose out of a recognition that workers with family responsibilities are disadvantaged by the demands of juggling 'familial' and waged labour. The Australian federal government announced the ratification of the convention in March 1990 and it came into effect on 30 March 1991.
24  Ironmonger and Sonius, 'Household productive activities'; M. Bittman, *Juggling Time: How Australian Families Use Time*, Office of the Status of Women, Department of Prime Minister and Cabinet, Canberra, 1991.
25  Office of the Status of Women, 'Working Families Issues Kit', Department of the Prime Minister and Cabinet, Commonwealth Government, AGPS, Fyshwick, ACTU, 1992.
26  M. Edwards, 'Commentary', in *Households Work*, ed. D. Ironmonger, Allen & Unwin, Sydney, 1989, op. cit., p. 35.
27  Goodnow, 'Work in households', p. 41.
28  Juliet Mitchell's approach summarised in R. W. Connell, *Gender and Power: Society, the Person and Sexual Politics*, Allen & Unwin, Sydney, 1987, pp. 96–98; see also M. Lake, Review (Connell), *Thesis Eleven*, no. 23, 1989, p. 162.
29  G. Rubin, 'The traffic in women: Notes on the 'political economy' of sex', in *Toward an Anthropology of Women*, ed. R. Reiter, Monthly Review Press, N.Y./London, 1975, p. 167.
30  Eldholm et al. cited in R. Rapp, 'Anthropology', *Signs*, vol. 4, no. 3, Spring 1979, pp. 507–8.
31  I have here reworked Frankel's critique of functionalist Marxist accounts of the relation between the economy and the 'superstructure'; B. Frankel,

*Marxian Theories of the State: A Critique of Orthodoxy*, Monograph
Series 3, Arena Pub., Melbourne, 1978.

32  The dangers of this approach are outlined in more detail in C. Beasley,
*Educating Rita's Grandmother: The Social Relation of the Sexes and
South Australian Curriculum Reform, 1875–1915*, unpublished M. Ed.
thesis, Flinders University of South Australia, 1984, pp. 19–37; see also
J. Allen, 'Marxism and the man question: Some implications of the
patriarchy debate', in *Beyond Marxism?: Interventions after Marx*, eds.
J. Allen and P. Patton, Intervention Pub., Sydney, 1983, pp. 91–111 and
C. Beasley, 'The patriarchy debate: Should we make use of the term
'patriarchy' in historical analysis?', *History of Education Review*, vol.
16, no. 2, 1987, pp. 13–20.

33  Connell, *Gender and Power*, pp. 96–8.

34  ibid., pp. 97 and 99–105; Lake, Review (Connell), pp. 163–163.

35  Goodnow, 'Work in households', p. 40

36  I have distinguished between the emotional labour provided in childcare
and other forms of 'care' because generally the latter can be flexibly
organised around a range of other activities including waged work.
However, care for aged relatives, for example, does sometimes demand
continuous supervision. The distinction is certainly a blurred one, but I,
in common with Delphy, am inclined to separate out labours specifically
directed towards the 'goal' of child-rearing. Delphy, 'Continuities and
discontinuities in marriage and divorce'.

37  No ready-made term describes the constant activation of (sexed) subjectivities that is an aspect of women's nurturing work in relation to
both children and, less commonly acknowledged, in relation to adults.
Nevertheless such activities, which run through women's talk within
households and elsewhere as well as through the range of labours
involved in caring, are an important element of the 'labour of love'
associated with domestic life. Women's link to emotionality and the
creation and upkeep of selves is not limited to their responsibility for
children but also occurs in relation to the activation/upkeep of the
masculinity/subjectivity of 'husbands'. Women's special responsibility for
'subjectivation' infuses 'the family' to such a degree that it is a particularly invisible form of labour.

38  J. Donzelot's approach noted in M. Barrett and M. McIntosh, *The
Anti-social Family*, Verso/NLB, London, 1982, p. 97. A. Ferguson's
account noted in S. Harding, 'The instability of the analytical categories
of feminist theory', *Signs*, vol. 11, no. 4, Summer 1986, pp. 648–9. H.
Hartmann, 'The historical roots of occupational segregation: Capitalism,
patriarchy and job segregation by sex', (Part II), *Signs*, vol. 1, no. 3,
Spring 1976, p. 139.

39  See, for example, work on heterosexuality and motherhood. A. Rich,
'Compulsory heterosexuality and lesbian existence', *Signs*, vol. 5, no. 4,
Summer 1980, pp. 631–690; A. Ferguson et al., 'On "Compulsory
heterosexuality and lesbian existence": Defining the issues', in *Feminist
Theory: A Critique of Ideology*, ed. N. Keohane et al., Harvester Press,
Brighton, Sussex, 1982, pp. 147–188; A. Rich, *Of Woman Born:
Motherhood as Experience and Institution*, Virago, London, 1983; M.

Hirsch, 'Mothers and daughters', *Signs*, vol. 7, no. 1, Autumn 1981, pp. 200–22.

40  C. MacKinnon, 'Feminism, Marxism, method and the state: An agenda for theory', *Signs*, vol. 7, no. 3, Spring 1982, passim; D. Dahlerup, 'Confusing concepts—confusing reality: A theoretical discussion of the patriarchal state', in *Women and the State: The Shifting Boundaries of public and Private*, ed. A. Sassoon, Hutchinson, London, 1987, pp. 93–127.

41  J. Allen, 'Does feminism need a theory of 'the state"?', in *Playing the State: Australian Feminist Interventions*, ed. S. Watson, Allen & Unwin, Sydney, 1990, pp. 21–37.

42  S. Harding, 'Why has the sex/gender system become visible only now?', in *Discovering Reality: Feminist Perspectives on Epistemology, Metaphysics, Methodology, and Philosophy of Science*, eds. S. Harding and M. Hintikka, D. Reidel Pub., Dordrecht, 1983, p. 311.

43  See Chapter 1, note 17.

44  R. Tong, *Feminist Thought: A Comprehensive Introduction*, Unwin Hyman, London, 1989, pp. 183, 189.

45  Harding, 'The instability of the analytical categories of feminist theory', p. 660; A. Ferguson, 'On conceiving motherhood and sexuality: A feminist materialist approach', in *Mothering: Essays in Feminist Theory*, ed. J. Trebilcot, Rowman & Allanheld, Totowa, N.J., 1984, p. 154.

46  Harding, 'The instability of the analytical categories of feminist theory', p. 660.

47  ibid.

48  The absolute (totalising) rejection of all forms of totality may be disputed. Such a rejection involves in the first instance a contradiction. Additionally, it can, for example, involve the obliteration of any perception of longstanding, generalised male domination and hence silence a possible mode of critique of the masculine domination of society and history. Although the category of totality is not innocent of authoritarian effects on knowledges, the utter repudiation of all understandings of forms of totality does not reinstate politically innocent categories. In other words, the term 'totality' requires careful deconstruction to investigate the possibility of different orders of analysis within it. I. Balbus, 'Disciplining women: Michel Foucault and the power of feminist discourse', *Praxis International*, vol. 5, no. 4, January 1986, p. 475.

49  Harding, 'The instability of the analytical categories of feminist theory', p. 660; b. hooks, *Feminist Theory from Margin to Center*, South End Press, Boston, Ma., 1983.

50  Harding, 'The instability of the analytical categories of feminist theory', p. 660.

## 3: Household work studies

1  T. Carlyle, 'The present time', *Latter Day Pamphlets*, 1, in J. M. and M. J. Cohen, *The Penguin Dictionary of Quotations*, Penguin, Harmondsworth, 1960, p. 99.

2  W. Shakespeare, *Love's Labour's Lost*, IV, iii, in ibid., p. 331.

3  For example, Ironmonger describes the household economy as motivated by 'altruism' without qualifying this term. D. Ironmonger, 'Households and the household economy', in *Households Work: Productive Activities, Women and Income in the Household Economy*, ed. D. Ironmonger, Allen & Unwin, Sydney, 1989, p. 3.

4  ibid., p. 4.

5  I use terms like 'household' and 'market' since they are commonly employed in the literature. Their use, nevertheless, requires some qualification, which is attempted in this chapter.

6  Cited in M. Bittman, *Juggling Time: How Australian Families Use Time*, Office of the Status of Women, Department of Prime Minister and Cabinet, Canberra, 1991, p. 1.

7  See Office of the Status of Women, *Selected Findings from Juggling Time*, Department of Prime Minister and Cabinet, Canberra, 1991, p. 2; Australian Institute of Family Studies data noted by J. Larkin and J. Brinkworth, 'Child rebate boost: New bid to pay wives', *The Advertiser*, March 25, 1991, p. 1.

8  Bittman, *Juggling Time*, p. 1.

9  ibid., pp. 31–32; D. Ironmonger, 'Preface', in *Households Work*, ed. D. Ironmonger, Allen & Unwin, Sydney, 1989, p. x.

10 E. Leacock, 'History, development, and the division of labour by sex: Implications for organization', *Signs*, vol. 7, no. 2, Winter 1981, p. 475. See also United Nations Division for Economic and Social Information, Department of Public Information, 'Worsening situation of women will be main issue confronting commission on the status of women', Note No. 22, International Women's Decade, February 13, 1980, p. 4. Information on Australia gathered by the Australian Bureau of Statistics (ABS), using two different methods for the reference years 1986–1987, estimated 'household production' as 57 per cent and 62 per cent of GDP. (ABS, 'Measuring unpaid household work: Issues and Experimental Estimates', ABS Information Paper, Catalogue No. 5236.0, Commonwealth Government, Canberra, 1990, p. 17.) See note 8 for similar 1990 estimate.

11 D. Ironmonger and E. Sonius, 'Household productive activities', in *Households Work*, ed. D. Ironmonger, Allen & Unwin, Sydney, 1989, p. 21; Leacock, 'History, development, and the division of labour by sex', p. 475.

12 ABS, 'Measuring unpaid household work', pp. 6–7.

13 Though Ironmonger expresses concern about the concentration on 'who does the work' in studies of the household economy, he recognises the importance of the issue. Ironmonger, 'Households and the household economy', p. 5; Ironmonger and Sonius, 'Household productive activities', pp. 20–1.

14 Ironmonger, 'Households and the household economy', p. 10.

15 J. Goodnow, 'Work in households: an overview and three studies', in *Households Work*, ed. D. Ironmonger, Allen & Unwin, Sydney, 1989, p. 43.

16 A. Giddens, *The Constitution of Society*, Polity Press, Cambridge, 1984, p. 25.

17  M. Rosaldo, 'The use and abuse of anthropology: Reflections on femi-
    nism and cross-cultural understanding', *Signs*, vol. 5, no. 3, Spring, 1980,
    p. 394.
18  Universal in the sense of eternal seems to me to raise problems different
    from those which occur when universal simply suggests a degree of
    historical continuity. Here I am employing the latter meaning.
19  A. Chadeau, *Measuring Household Activities: Some International Com-
    parisons*, International Association for Research in Income and Wealth,
    presented at the 18th General Conference of the Association, Luxem-
    bourg, August 1983, cited in ABS, 'Measuring unpaid household work',
    pp. 5 and 7.
20  C. Delphy, 'A materialist feminism is possible', *Feminist Review*, no. 4,
    1980, p. 87.
21  A. Gouldner, *The Coming Crisis of Western Sociology*, Basic Books,
    N.Y., 1970, p. 35.

    [The theorist] must have the courage of his convictions, or at least
    courage enough to acknowledge his beliefs as his, whether or not
    legitimated by reason and evidence. Unless he delivers his domain
    assumptions from the dim realm of subsidiary awareness into the clearer
    realm of focal awareness, where they can be held firmly in view, they
    can never be brought before the bar of reason or submitted to the
    evidence.

22  Goodnow, 'Work in households', p. 44; M. Edwards, 'Individual equity
    and social policy', in *Women, Social Science and Public Policy*, eds. J.
    Goodnow and C. Pateman, Allen & Unwin, Sydney, 1985, pp. 95–103;
    H. Hartmann, 'The family as the locus of gender, class and political
    struggle: The example of housework', in *Feminism and Methodology:
    Social Science Issues*, ed. S. Harding, Indiana University Press/Open
    Press, Bloomington, Indiana/Milton Keynes, 1987, pp. 112 and 117–29.
23  Ironmonger, 'Households and the household economy', pp. 6–7; Iron-
    monger and Sonius, 'Household productive activities', pp. 24–5; ABS,
    'Measuring unpaid household work', pp. 1–3.
24  R. Tong, *Feminist Thought: A Comprehensive Introduction*, Unwin
    Hyman, London, 1989, p. 135 (Elshtain's approach noted in relation to
    her discussion of Susan Brownmiller); J. Elshtain, *Public Man, Private
    Woman*, Princeton University Press, Princeton, 1981.
25  J. Allen, Review (Hartsock), *Signs*, 10:3, Spring 1985, p. 578;
26  E. Garnsey, 'Women's work and theories of class and stratification',
    noted in M. Huxley, 'Commentary' in *Households Work*, ed. D. Iron-
    monger, Allen & Unwin, Sydney, 1989, p.81; C. Delphy, *Close to Home:
    A Materialist Analysis of Women's Oppression*, trans. and ed. D. Leon-
    ard, Hutchinson, London, 1984.
27  A. Jamrozik, 'The household economy and social class', in *Households
    Work*, ed. D. Ironmonger, Allen & Unwin, Sydney, 1989, p. 75.
28  Ironmonger, 'Households and the household economy', p. 10.
29  This point regarding weak connections between household and market
    economies is made in the same book in which Ironmonger and Jamrozik
    present their approaches, that is, Ironmonger ed., *Households Work*, See
    Goodnow, 'Work in households', p. 45.

30  Bittman, *Juggling Time*, pp. 3, 9, 21, 22 and 46; C. O'Donnell and P. Hall, *Getting Equal*, Allen & Unwin, Sydney, 1988, p. 22; G. Russell, *The Changing Role of Fathers*, University of Queensland Press, Brisbane, 1983, p. 53; M. Sacks, 'Unchanging times: A comparison of the everyday life of Soviet working men and women between 1923 and 1966', *Journal of Marriage and the Family*, vol. 39, November 1977, pp. 793–805.
31  Goodnow, 'Work in households', p. 45.
32  O. Hawrylyshyn, 'Estimating the value of household work in Canada', 1971, quoted in ABS, 'Measuring unpaid household work', p. 1.
33  M. Murphy, 'Comparative estimates of the value of household work in the United States for 1976', Review of Income and Wealth, Series 28, 1982, quoted in ABS, 'Measuring unpaid household work', p. 1.
34  However, as has been noted, some labours which are marketable, such as sexual activities, are oddly omitted from these studies.
35  S. Lewenhak, *The Revaluation of Women's Work*, second edition, Earthscan Pub., London, 1992, p. 9.
36  ABS, 'Measuring unpaid household work', p. 17. See Ferber's discussion of the limitations of neoclassical theories in dealing with the private sphere. M. Ferber, 'Women and work: Issues of the 1980s', *Signs*, vol. 8, no. 2, Winter 1982, pp. 273–295; E. McCrate, 'Comment on Ferber's 'Women and work: Issues of the 1980s'', *Signs*, vol. 9, no. 2, Winter 1983, pp. 326–30.
37  ABS, 'Measuring unpaid household work', p. 7.
38  M. Edwards, 'Commentary', *Households Work*, p. 34.
39  Tong, *Feminist Thought*, pp. 31–38 and 61–69; Elshtain, *Public Man, Private Woman*; J. Elshtain, *Meditations on Modern Political Thought: Masculine/Feminine Themes from Luther to Arendt*, Praeger, N.Y., 1986; A. Jaggar, *Feminist Politics and Human Nature*, Rowman & Allanheld, Totowa, N.J., 1983.
40  M. Campioni and E. Gross, 'Love's labours lost: Marxism and feminism', in *Beyond Marxism?: Interventions After Marx*, eds. J. Allen and P. Patton, Intervention Pub., Sydney, 1983, pp. 123 and 133–4.
41  V. Seidler, *Recreating Sexual Politics: Men, Feminism and Politics*, Routledge, London, 1991, p. 177.

## 4 Towards a 'materialist' feminism?

1  L. Carroll, *Through the Looking Glass*, in *The Oxford Dictonary of Quotations*, Second edition Oxford University Press, London, 1953, p. 131.
2  R. Tong, *Feminist Thought: A Comprehensive Introduction*, Unwin Hyman, London, 1989, p. 175.
3  M. Fonow, Review (Jaggar, Vogel, Nicholson), *Signs*, vol. 13, no. 2, Winter 1988, pp. 349–350. See L. Vogel, *Marxism and the Oppression of Women: Toward a Unitary Theory*, Rutgers University Press, New Brunswick, N.J., 1984.
4  H. Hartmann, 'The historical roots of occupational segregation: Capitalism, patriarchy and job segregation by sex', (Part II), *Signs*, vol. 1, no. 3, Spring 1976, p. 167.

5 For one account of the problems of the Marxist focus on the 'market', see B. Theile, 'Vanishing acts in social and political thought: Tricks of the trade', in *Feminist Challenges: Social and Political Theory*, eds. C. Pateman and E. Gross, Allen & Unwin, Sydney, 1986, pp. 34 and 43.

6 D. Ironmonger and E. Sonius, 'Household productive activities', in *Households Work: Productive Activities, Women and Income in the Household Economy*, ed. D. Ironmonger, Allen & Unwin, Sydney, 1989, p. 21; A. Jamrozik, 'The household economy and social class', in *Households Work*, ed. D. Ironmonger, Allen & Unwin, Sydney, 1989, p. 75; P. Thompson, *The Nature of Work: An Introduction to Debates on the Labour Process*, Macmillan, London, 1983, pp. 199–201 and 208.

7 Tong, *Feminist Thought*, pp. 63–65 and 182; Fonow, Review (Jaggar, Vogel, Nicholson), pp. 348–9; I. Young, 'Beyond the unhappy marriage: A critique of the dual systems theory', in *Women and Revolution: A Discussion of the unhappy marriage of Marxism and Feminism*, ed. L. Sargent, South End Press, Boston, 1981, pp. 43–69.

8 Young, 'Beyond the unhappy marriage', pp. 52–6; I. Young, 'Is male gender identity the cause of male domination?', in *Mothering: Essays in Feminist Theory*, ed. J. Trebilcot, Rowman & Allanheld, Totowa, N.J., 1984, pp. 137–8, 140–1 and 143–4.

9 Young, 'Is male gender identity the cause of male domination?', pp. 138–9.

10 ibid., pp. 143–144; Young, 'Beyond the unhappy marriage', pp. 54–6.

11 Young, 'Is male gender identity the cause of male domination?', pp. 138–141; Young, 'Beyond the unhappy marriage, pp. 54–5.

12 G. Rubin, 'The traffic in women: Notes on the 'political economy' of sex', in *Toward an Anthropology of Women*, ed. R. Reiter, Monthly Review Press, N.Y./London, 1975, p. 167; A. Ferguson, 'On conceiving motherhood and sexuality: A feminist materialist approach', in *Mothering: Essays in Feminist Theory*, ed. J. Trebilcot, Rowman & Allanheld, Totowa, N.J., 1984, p. 156.

13 Tong, *Feminist Thought*, p. 182; Young, 'Beyond the unhappy marriage', pp. 58–61.

14 Tong, *Feminist Thought*, p. 182. Marxist and socialist feminists among others commonly argue that capitalism *created* the split between the public ('economy') and the private ('the family'); see for example Nicholson and Zaretsky. L. Nicholson, *Gender and History: The Limits of Social Theory in the Age of the Family*, Columbia University Press, N.Y., 1986; J. Butler, 'Gender, the family and history', *Praxis International*, 7:1, April 1987, p. 128; E. Zaretsky, 'Socialism and feminism I: Capitalism, the family, and personal life, Part I', *Socialist Revolution*, vol. 3, nos. 1/2, January–April 1973, passim.

15 J. Allen, 'Marxism and the man question: Some implications of the Patriarchy Debate', in *Beyond Marxism?*, eds. J. Allen and P. Patton, Intervention Pub., Sydney, 1983, p. 104 and endnote 53, p. 111. Similarly Allen argues (p. 105) that capitalism did not create the dependent married woman.

16 ibid., pp. 104–5.

17  J. Jacquette, 'Power as ideology: A feminist analysis', in *Women's Views of the Political World of Men*, ed. J. Stiehm, Transnational Pub., N.Y., p. 23. In noting that the political experience of women—that is, their relation to or use of power, may be affected by a public–private divide, feminist writers such as Jacquette and Markus have pondered the degree to which politics in the private realm is distinct from that of the public sphere and the extent to which forms of power may be capable of bridging the distinction. Young does not seem to recognise these issues, despite the possible effect they might have on her account of the interconnection between capitalism and the sexual organisation of labour. M. Markus, '"Deconstructed" inequality: Women and the public sphere', *Thesis Eleven*, No. 14, 1986, p. 127.

18  England and Farkas's approach noted in J. Goodnow, 'Work in households: an overview and three studies', *Households Work*, ed. D. Ironmonger, Allen & Unwin, Sydney, 1989, p. 45.

19  Thompson, *The Nature of Work*, p. 40.

20  Jaggar privileges use of a Marxist category defined by its relation to labour and the products of labour, despite her recognition of the unique character of women's experience and position and her acceptance that patriarchy only incompletely intersects with capitalism. Tong, *Feminist Thought*, pp. 186–189 and 65; Fonow, Review (Jaggar, Vogel, Nicholson), pp. 348–349.

21  Editorial, *Feminist Review*, no. 23, June 1986, p. 5.

22  Young, 'Beyond the unhappy marriage', pp. 58-62.

23  Barrett, for example, has noted a reductive circularity in the work of dual systems theorists such as Zillah Eisenstein. Reductions of capitalism (and colonialism) to patriarchy are also evident in the approach of radical feminists like Azizah Al-Hibri. M. Barrett, *Women's Oppression Today: Problems in Marxist Feminist Analysis*, first edition, Verso/NLB, London, 1980, p. 16; Z. Eisenstein, 'Developing a theory of capitalist patriarchy and socialist feminism', in *Capitalist Patriarchy and the Case for Socialist Feminism*, ed. Z. Eisenstein, Monthly Review Press, N.Y./London, 1979, p. 5; A. Al-Hibri, 'Reproduction, mothering, and the origins of patriarchy', in *Mothering*, ed. J. Trebilcot, Rowman & Allanheld, Totowa, N.J., 1984, pp. 83 and 90. For one discussion of the difficulties of reducing patriarchy to capitalism or vice versa, see M. Campioni and E. Gross, 'Love's labours lost: Marxism and feminism', in *Beyond Marxism?*, eds. J. Allen and P. Patton, Intervention Pub., Sydney, 1983, p. 138.

24  Editorial, *Feminist Review*, p. 7; A. Curthoys, 'Reply to Rosemary Pringle', *Australian Feminist Studies*, nos. 7/8, Summer 1988, pp. 175 and 177; A. Edwards, *Regulation and Repression: The Study of Social Control*, Allen & Unwin, Sydney, 1988, pp. 52, 55 and 56.

25  Curthoys, 'Reply to Rosemary Pringle', pp. 175 and 172.

26  Barrett, *Women's Oppression Today*, first edition, p. 249; Allen, 'Marxism and the Man Question', p. 95.

27  Campioni and Gross, 'Love's labours lost, p. 138.

28  C. Beasley, 'The patriarchy debate: Should we make use of the term 'patriarchy' in historical analysis?', *History of Education Review*, vol.

16, no. 2, 1987, p. 19; Campioni and Gross, 'Love's labours lost, pp. 116–117; C. Delphy, 'A materialist feminism is possible', *Feminist Review*, no. 4, 1980, pp. 95–104; Allen, 'Marxism and the Man Question', pp. 94–6 and 105–6.

29 Barrett's revised edition of *Women's Oppression Today*, published eight years after the first edition, includes a new introduction which reassesses many of her earlier views in the light of new developments in feminist and other theories. However, important aspects of the earlier perspective I have criticised remain. By contrast with Althusser, she argues that a central theme in her book was to propose that links between class and 'gender' should be seen historically, rather than in functional terms (pp. xvii, xiv–xv, xviii and xvi). She does not appear to notice, as Judith Allen's commentary on her work makes clear, that her approach simply replaced the abstracted functionalism employed by Althusser with a historical functionalism. Barrett's approach retains a functionalist link between class and 'gender' because her account of history rests upon the transposition of a Marxist/class-based understanding of 'history' onto the latter. Moreover, there is little evidence in the introduction to the second edition of any marked rethinking of the 'ideological' view of sex relations outlined in the first edition. Since Barrett assumes that the way to deal with the question of materialism/economics in the field of 'gender' is to return to the touchstone of 'history' (pp. xvi–xvii), it would seem that there is not much space here for any notion of a specific or distinct sexual economy. She conceives 'gender', after all, as historically embedded in class relations (p. 249) and sees history itself through the lens of Marxist 'modes of production', as her analysis of the term 'patriarchy' demonstrates (pp. 14, 15 and 19). M. Barrett, *Women's Oppression Today: The Marxist/Feminist Encounter*, second edition, Verso/NLB, London, 1988; Allen, 'Marxism and the man question', pp. 95 and 100–1.

30 Barrett *implicitly* replicates Mitchell's ideological account of sex relations; Tong, *Feminist Thought*, pp. 175–6.

31 Much of Mitchell's analysis *could*, however, be conceived in terms of labour.

32 Elshtain, *Public Man, Private Woman*, pp. 243 and 265.

33 R. Rowland, 'Reproductive technologies: The final solution to the woman question', in *Test-Tube Women: What Future for Motherhood?*, eds. R. Arditti et al., Pandora Press, London/Boston, 1984, p. 369.

34 C. MacKinnon, 'Feminism, marxism, method, and the state: An agenda for theory', *Signs*, vol. 7, no. 3, Spring 1982, pp. 524–7; C. MacKinnon, 'Reply to Miller, Acker and Barry, Johnson, West, and Gardiner', *Signs*, vol. 10, no. 1, Autumn 1984, pp. 184–5.

35 MacKinnon, 'Reply to Miller, Acker and Barry, Johnson, West, and Gardiner', p. 185.

36 Here I disagree with Tong's view that 'unified system' theorists such as Young and Jaggar cannot be criticised for their use of Marxism or for their concentration on women as workers. Even though Young and Jaggar do not reduce women's labour to waged labour, they continue to reduce 'the woman question' to a production paradigm, to an

enhanced account of 'the worker question'. Tong, *Feminist Thought*, p. 189.

37  O'Brien argues that feminism must transcend 'male–stream' theory. M. O'Brien, *The Politics of Reproduction*, RKP, London/Boston, 1981, p. 6.

38  Jaggar's approach summarised in Fonow, Review (Jaggar, Vogel, Nicholson), p. 349.

39  This, in Bland et al. and Delphy's terms, is necessary even for an understanding of women's position in *waged* labour. L. Bland et al., 'Women 'inside' and 'outside' the relations of production', in *Women Take Issue: Aspects of Women's Subordination*, eds. Centre for Contemporary Cultural Studies (Women's Studies Group), Hutchinson/CCCS, London, 1978, p. 35; C. Delphy, 'Continuities and discontinuities in marriage and divorce', in *Sexual Division and Society: Process and Change*, eds. D. Barker and S. Allen, Tavistock, London, 1976, p. 79.

40  M. Edwards, 'Commentary', in *Households Work*, ed. D. Ironmonger, Allen & Unwin, Sydney, 1989, p. 34.

41  I will discuss this focus and terminology further in Chapters 5 and 6.

42  N. Hartsock, 'The feminist standpoint: Developing the ground for a specifically feminist historical materialism', in *Feminism and Methodology: Social Science Issues*, ed. S. Harding, Indiana University Press/Open University Press, Bloomington, Indiana/Milton Keynes, 1987, p. 164. See her examples pp. 165–7 and pp. 170–1.

43  Ironmonger and Sonius, 'Household productive activities', p. 18; Goodnow, 'Work in households', p. 41.

44  See for summary, J. Allen, Review (Hartsock), *Signs*, vol. 10, no. 3, Spring 1985, p. 578.

45  S. Harding, 'Conclusion: Epistemological questions', in *Feminism and Methodology, ed. S. Harding, Indiana University Press/Open University Press, Bloomington, Indiana/Milton Keynes, 1987, p. 187.*

46  Millman and R. Kanter, 'Introduction to *Another Voice: Feminist Perspectives on Social Life and Social Science*', in *Feminism and Methodology, ed. S. Harding, Indiana University Press/Open University Press, Bloomington, Indiana/Milton Keynes, 1987, pp. 31–5.*

47  K. Mumford, *Women Working: Economics and reality*, Allen & Unwin, Sydney, 1989.

48  R. Sharp and R. Broomhill, *Short-changed: Women and economic policies*, Allen & Unwin, Sydney, 1988, pp. 167–9.

49  M. Waring, *Counting for Nothing: What Men Value and What Women Are Worth*, Allen & Unwin/Port Nicholson Press, Sydney/Wellington, 1988, p. 227. Waring's unequivocal stance in considering women's labour is that 'production'/labour in the market is no different from that in the private sphere.

> 'Goldschmidt-Clermont argues that the transformation of natural resources(for example, agricultural products into goods . . .) . . . does not change in nature when passing the border line of monetary exchanges . . . While it is clear that the methods evaluated by Goldschmidt-Clermont have been used in the *developed* world, and on

*housework*, it is quite clear to me that they can, and should, be developed to encompass *all* unpaid work that women do (p. 226).

There is no awareness evident here that not all unpaid work involves the transformation of natural 'resources' in the sense of physical objects or their transmutation into 'goods', let alone any uncertainty regarding monetary measures of the 'value' of unpaid work.

50   Harding, 'Conclusion: Epistemological questions', p. 189.
51   A. Phillips, *Hidden Hands: Women and Economic Policies*, Pluto Press, London, 1983, p. 34.
52   E. Gross, 'Conclusion: What is feminist theory?', in *Feminist Challenges*, eds. C. Pateman and E. Gross, Allen & Unwin, Sydney, 1986, p. 197.
53   Similar points are raised by Hartsock and Ruddick. Hartsock, 'The feminist standpoint', p. 175; S. Ruddick, 'Maternal thinking' in *Mothering*, ed. J. Trebilcot, Rowman & Allanheld, Totowa, N.J., 1984, pp. 226–7; S. Ruddick, 'Preservative love and military destruction: Some reflections on mothering and peace', in *Mothering*, ed. J. Trebilcot, Rowman & Allanheld, Totowa, N.J., 1984, p. 239.
54   Studies by Walker and by Sacks noted in H. Hartmann, 'The family as the locus of gender, class and political struggle: The example of housework', in *Feminism and Methodology*, ed. S. Harding, Indiana University Press/Open University Press, Bloomington, Indiana/Milton Keynes, 1987, p. 126.
55   Sacks noted in Hartmann, 'The Family as the Locus of Gender, Class and Political Struggle', p. 126.
56   I am indebted here to MacKinnon's account of the meaning of the term 'agenda' and her understanding of 'toward'/'towards', which influenced the title of this chapter. MacKinnon, 'Reply to Miller, Acker and Barry, Johnson, West, and Gardiner', p. 184.

## 5 Dual/multiple vision

1   T. Tusser, 'Preface to the Book of Housewifery', *Five Hundred Points of Good Husbandry*, quoted in J. M. and M. J. Cohen, *The Penguin Dictionary of Quotations*, Penguin, Harmondsworth, 1960, p. 400.
2   R. Bradbury, 'The Last Circus', in R. Bradbury, *The Toynbee Convector*, Grafton Books, London, 1988, p. 82.
3   It is possible that sexual features of women's *waged* work (such as partially or non-commodified elements related to care/nurturing) could be regarded as an 'informal' sector of the 'household' economy, in much the same way that outwork is a market activity—albeit performed in the home—and hence is seen as an 'informal' sector of the market economy.
4   J. Cocks quoted in R. Tong, *Feminist Thought: A Comprehensive Introduction*, Unwin Hyman, London, 1989, p. 131.
5   See for example some possible concerns regarding postmodernism/poststructuralism raised by Hartsock and a broad summary of various feminist critiques in Heckman. N. Hartsock, 'Foucault on power: A theory for women?', in *Feminism/Postmodernism*, ed. L. Nicholson, Routledge, N.Y., 1990, pp. 157–175; S. Heckman, *Gender and Knowl-*

*edge: Elements of a Postmodern Feminism*, Polity Press, Cambridge, 1990, pp. 152–190.

6  b. hooks, *Feminist Theory from Margin To Center*, South End Press, Boston, 1983, pp. 43–4. Since Gloria Watkins spells her pseudonym, bell hooks, without capitals I have followed her example.

7  hooks in both her early and more recent works displays a deeply-felt antagonism to what she views as the class interested nature of bourgeois/liberal feminism and links much of her critique of the assumption of commonality to her account of the limits of this form of feminism. However, her more recent writings suggest that she is not necessarily antagonistic to any and all conceptions of connected, if not shared, experiences in relation to sexual positioning. She notes approvingly, in *Talking Back*, Radford-Hill's criticism of work by black women producing feminist theory (which includes criticism of hooks's own work). Radford-Hill sees these women as focusing more on white women's racism and 'the importance of race difference, than on the ways in which feminist struggle could strengthen and help black communities'. hooks declares that this latter issue is her 'current concern'. hooks goes on to reject any notion of feminist struggle as 'a white female thing that has nothing to do with black women', and notes that '[c]urrent feminist scholarship can be useful to black women, particularly work on parenting'. hooks cites the work of Dorothy Dinnerstein as an example of such scholarship. This outlook implies at least some recognition of aspects of commonality in relation to women's labours. b. hooks, *Feminist Theory* pp. 1–15; b. hooks, *Yearning: Race, Gender, and Cultural Politics*, South End Press, Boston, Mass., 1990, pp. 220-1; b. hooks, *Talking Back*, South End Press, Boston, Mass., 1989, pp. 178–9 and 182.

8  Hartmann points out that the evidence from studies of household work suggests many consistencies in women's private labour, and that the limited data available on time usage, at least, indicates little variation in women's experience across class and race differences. The work of Bittman and of Russell in Australia supports this general point in that their studies show consistent patterns of women's comparatively unequal commitment to domestic labour regardless of class variables. Hartmann, 'The family as the locus of gender, class, and political struggle: The example of housework', in *Feminism and Methodology: Social Science Issues*, ed. S. Harding, Indiana University Press/Open University Press, Bloomington, Indiana/Milton Keynes, 1987, p. 123; M. Bittman, *Juggling Time: How Australian Families Use Time*, Office of the Status of Women, Department of Prime Minister and Cabinet, Canberra, 1991, pp. 31–2; G. Russell, *The Changing Role of Fathers*, University of Queensland Press, Brisbane, 1983, pp. 67 and 73.

9  N. Cott, quoted in Mascia-Lees et al., 'The postmodernist turn in anthropology: Cautions from a feminist perspective', *Signs*, vol. 15, no. 1, Autumn 1989, p. 27.

10  S. Harding, 'The Instability Of The Analytical Categories Of Feminist Theory', *Signs*, vol. 11, no. 4, Summer 1986, p. 660.

11  See related note, Chapter 2, note 46. 'Totality'/'totalities' may be con-
    ceived as limited in scope, multiple and heterogeneous.
12  S. Harding, 'Conclusion: Epistemological questions', in *Feminism and
    Methodology*, ed. S. Harding, Indiana University Press/Open University
    Press, Bloomington, Indiana/Milton Keynes, 1987, p. 189.
13  Tong, *Feminist Thought*, p. 181.
14  M. Campioni and E. Gross, 'Love's labours lost: Marxism and feminism'
    in *Beyond Marxism: Interventions After Marx*, eds. J. Allen and P.
    Patton, Intervention Pub., Sydney, 1983, pp. 116–7.
15  Z. Eisenstein, 'Developing a theory of capitalist patriarchy and socialist
    feminism', in *Capitalist Patriarchy and the Case for Socialist Feminism*,
    ed. Z. Eisenstein, Monthly Review Press, N.Y./London, 1979, p. 5.
16  N. Chodorow, 'Mothering, male dominance and capitalism', in *Capitalist
    Patriarchy and the Case for Socialist Feminism*, ed. Z. Eisenstein,
    Monthly Review Press, N.Y./London, 1979, p. 100.
17  I. Young, 'Is male gender identity the cause of male domination?', in
    *Mothering: Essays in Feminist Theory*, ed. J. Trebilcot, Rowman &
    Allanheld, Totowa, N.J., 1984, p. 129. Similarly I disagree with
    Nicholson's assertion that Chodorow straightforwardly views the social
    as created and explained by the psychic. J. Butler, 'Gender, the family
    and history', *Praxis International*, vol. 7, no. 1, April 1987, p. 127.
18  The account of an economic system for sex relations that Chodorow
    does provide is in any case largely restricted to the sexual division of
    labour in childcare because of her belief in the centrality of this arena
    to sexual hierarchy. N. Chodorow, *The Reproduction of Mothering:
    Psychoanalysis and the Sociology of Gender*, University of California
    Press, Berkeley/Los Angeles, 1978.
19  S. Harding, 'Why has the sex/gender system become visible only now?',
    in *Discovering Reality: Feminist Perspectives on Epistemology, Meta-
    physics, Methodology, and Philosophy of Science*, eds. S. Harding and
    M. Hintikka, D. Reidel Pub., Dordrecht, 1983, p. 311.
20  Eisenstein, 'Developing a theory of capitalist patriarchy and socialist
    feminism', pp. 27–8.
21  Harding, 'Conclusion: Epistemological questions', p. 185. See also this
    chapter, note 38, for a further discussion of problems in Harding's
    account of the 'economic'.
22  It should be noted that some writers in the field of household work
    studies have made use of the work of Delphy and Hartmann, for
    example, and therefore the approaches in these studies are not entirely
    distinct from those of the two feminists cited. See, for instance, J.
    Goodnow, 'Work in households: an overview and three studies' and M.
    Huxley, 'Commentary', in *Households Work: Productive activities,
    women and income in the household economy*, ed. D. Ironmonger, Allen
    & Unwin, Sydney, 1989, pp. 39 and 44 and pp 80-1.
23  While Hartmann, Ferguson and Delphy clearly deal with conceptions of
    dual/multiple materiality, Hartsock has proposed a feminist standpoint
    epistemology which Harding views as evading the problems of 'dual
    systems' theory and opening the way for a multisystem model. Harding,
    'The instability of the analytical categories of feminist theory', p. 660.

24  G. Rubin, 'The traffic in women: Notes on the 'Political Economy' of sex', in *Toward an Anthropology of Women*, ed. R. Reiter, Monthly Review Press, N.Y./London, 1975, pp. 168 and 159.
25  C. Delphy, 'Continuities and discontinuities in marriage and divorce' and S. Allen and D. Barker, 'Sexual divisions and Society, in *Sexual Divisions and Society: Process and Change*, eds. D. Barker and S. Allen Tavistock, London, 1976, pp. 77–8 and pp. 5–7.
26  Campioni and Gross, 'Love's labours lost', p. 138.
27  Delphy, 'Continuities and discontinuities in marriage and divorce', p. 78; pp. 84 and 86–7.
28  ibid., p. 79.
29  R. W. Conneil, *Gender and Power: Society, the Person and Sexual Politics*, Allen & Unwin, Sydney, 1987, p. 105.
30  ibid., p. 104.
31  J. Allen, 'Marxism and the Man Question: Some Implications of the Patriarchy Debate' in *Beyond Marxism*, eds. J. Allen and P. Patton, Intervention Pub., Sydney, 1983, p. 94. See also Campioni and Gross' related account of their discomfort with perspectives which conflate patriarchy with capitalism.

> More sophisticated versions of socialist feminism now present capitalism as the most advanced state (or the most oppressive version) of patriarchy. Though this may seem perilously close to the radical feminist view, we would argue that it co-opts radical feminism for two reasons. Firstly, it conveniently displaces the problem of patriarchy once more from the question of male domination to its apparent expression in social reality: capitalism. Secondly, it makes it virtually impossible not to combat capitalism as the *primary* target, if indeed it is the most oppressive version of patriarchy.

Campioni and Gross, 'Love's labours lost', p. 138.
32  Australian Bureau of Statistics (ABS), 'Measuring unpaid household work: Issues and experimental estimates', ABS Information Paper, Catalogue No. 5236.0, Commonwealth Government, Canberra, 1990,     p. 17.
33  A. Ferguson, 'On conceiving motherhood and sexuality: A feminist materialist approach', in *Mothering: Essays in Feminist Theory*, ed. J. Trebilcot, Rowman & Allanheld, Totowa, N.J., 1984, p. 154–155 and 176. It could also be argued that Hartmann and Hartsock move beyond a dual model to *some* extent in so far as the former occasionally notes race/ethnicity as an element of social relations which interacts with patriarchy and the latter mounts a standpoint approach intended to allow for multiplicity. H. Hartmann, 'The unhappy marriage of Marxism and feminism: Towards a more progressive union', in *Women and Revolution: A Discussion of the Unhappy Marriage of Marxism and Feminism*, ed. L. Sargent, South End Press, Boston, 1981, for example, pp. 14–15; Hartsock, see this chapter, note 23.
34  Ferguson, 'On Conceiving Motherhood and Sexuality', pp. 154–7.
35  H. Hartmann, 'The historical roots of occupational segregation: Capitalism, Patriarchy, and Job Segregation by Sex', (Part II), *Signs*, vol. 1, no. 3, 1976, pp. 147–169; N. Hartsock, 'The feminist standpoint:

Developing the ground for a specifically feminist historical materialism',
in *Feminism and Methodology*, ed. S. Harding, Indiana University
Press/Open University Press, Bloomington, Indiana/Milton Keynes, 1987,
pp. 157–180.

36 Ferguson refers to two semi-autonomous economies, 'one centred in the
household and one in capitalist production', both of which she views as
'gendered' and sexualised. Nevertheless, she conceives the mode of
'sex/affective production' linked to the household as 'organising the
sex/affective aspects of the work bonds of the workforce' in the capitalist
economic system. A. Ferguson, *Blood at the Root: Motherhood, Sexuality and Male Dominance*, Pandora Press, London, 1989, p. 78 and
83.

37 Hartmann, 'The unhappy marriage of Marxism and feminism', pp. 15
and 16.

38 Harding, for example, notes Hartmann's acceptance of aspects of Marxism, which, she asserts, undercuts Hartmann's account of materialism in
patriarchy. Harding, correctly in my view, castigates Hartmann for failing
to include processes related to the development of sexual identity in the
latter's analysis of materialism within patriarchy. Nevertheless, as pointed
out earlier, Harding's own approach seems debatable. Harding sees
processes connected with sexual identity as 'material', but oddly enough
not as 'economic'. Since she appears to be aware that sexual identity is
a 'production', it is unclear why the constitution of identity is only
conceived as an illustration of the materiality of the psychological rather
than as also involving labour and thus as part of the economic organisation of sex relations. S. Harding, 'What is the real material base of
patriarchy and capital?', in *Women and Revolution*, ed. L. Sargent, South
End Press, Boston, 1981, pp. 142–7.

39 Hartmann, 'The unhappy marriage of Marxism and feminism', pp. 3
and 2.

40 ibid., p. 17

## 6 Conceiving a sexual epistemology of economics

1 U. Eco, *Foucault's Pendulum*, Picador, London, 1989, pp. 365 and 469.

2 J. Fowles, *The Magus*, in J. M. and M. J. Cohen, *The Penguin Dictionary
of Modern Quotations*, second edition, Penguin, Harmondsworth, 1980,
p. 121.

3 K. Salleh, 'Contribution to the critique of political epistemology', *Thesis
Eleven*, no. 8, January 1984, p. 29.

4 H. Hartmann, 'The unhappy marriage of Marxism and feminism:
Towards a more progressive union', in *Women and Revolution: A
discussion of the unhappy marriage of Marxism and Feminism*, ed. L.
Sargent, South End Press, Boston, 1981, p. 15.

5 A. Ferguson, 'On conceiving motherhood and sexuality: A feminist
materialist approach', in *Mothering: Essays in Feminist Theory*, ed. J.
Trebilcot, Rowman & Allanheld, Totowa, N.J., 1984, pp. 153–9.

6 N. Hartsock, 'The feminist standpoint: Developing the ground for a
specifically feminist historical materialism', in *Feminism and Methodol-*

ogy: *Social Science Issues*, ed. S. Harding, Indiana University Press/Open University Press, Bloomington, Indiana/Milton Keynes, 1987, p. 175.

7  C. Delphy, 'A materialist feminism is possible', *Feminist Review*, no. 4, 1980, p. 92.

8  Both Tong and Harding have noted Hartmann's largely unaltered use of Marx's account of capitalism and economic categories derived from this. There are few analyses of Delphy's work, but some, like that of Thompson, have drawn attention to her inclination to accept Marxist categories developed in relation to waged labour, which supports my view of her seemingly uncritical stance with regard to Marx's account of capitalism. R. Tong, *Feminist Thought: A Comprehensive Introduction*, Unwin Hyman, London, 1989, p. 180; P. Thompson, *The Nature of Work: An Introduction to Debates on the Labour Process*, Macmillan, London, 1983, p. 265.

9  Delphy, 'A materialist feminism is possible', p. 92 (sexuality) and pp. 95–8 (ideology and psychology). Thompson suggests that Delphy 'marginalises' issues like sexuality because she makes the mistake of applying Marxist categories derived from the market economy to the private sphere and describes that sphere in 'rigidly economistic terms'. I would go further. Delphy not only inappropriately applies market categories to a qualitatively and quantitatively different economy, but she also assumes that Marxian analysis of 'the economic' realm is equivalent to economic processes per se (Marxian *method* is regarded as describing economics in general) and adopts a very narrow 'economistic' version of Marxism at that. It is not just that Delphy's use of Marxian categories is economistic, because these are applied to sex relations, as Thompson proposes. This use is based on an economistic reading of Marxism and is not necessarily appropriate even to class relations.
   Delphy's version of Marxism does reflect a tendency within that tradition's definition of materiality to exclude libidinal, linguistic and 'psychological' processes and to ignore an element of labour in them. However, she provides a very orthodox reading of the Marxist account of materiality which would not be accepted by at least some Marxists. The main point here is that Delphy's economistic omission of 'psychology', sexuality and ideology replicates problems in Marxism (perhaps showing the limits of Marxism in even dealing with capitalist economics) that are certainly serious when one is dealing with the economics of sex relations. Thompson, *The Nature of Work* p. 265; M. Campioni and E. Gross, 'Love's labours lost: Marxism and feminism' in *Beyond Marxism: Interventions after Marx*, eds. J. Allen and P. Patton, Intervention Pub., Sydney, 1983, p. 129; S. Harding, 'What is the real material base of patriarchy and capital?', in *Women and Revolution*, ed. L. Sargent, South End Press, Boston, 1981, pp. 144–5.

10  This is another way of taking up Juliet Mitchell's point that sexual subjectivities and the constitution of the sexual order are buried *very* deep in social relations; in her terms, deep in the psyche. See Tong's summary account of Mitchell's approach. Tong, *Feminist Thought*, p. 171.

11 Hartsock, 'The feminist standpoint', p. 166. Both Harding and Ferguson
   note in different ways an insufficiently recognised *dissimilarity* between
   the production of things and the production of people which has
   marginalised understanding of aspects of private labour as against waged
   labour. Harding, 'What is the real material base of patriarchy and
   capital?', pp. 145–6; Ferguson, 'On conceiving motherhood and
   sexuality', pp. 155–6.
12 Delphy, 'A materialist feminism is possible', p. 96, 98.. Delphy declares
   that ideology is not restricted to subjectivity, but it should also be said
   that subjectivity is not only 'superstructural'. Subjectivity is implicated
   in the economics of sex relations: sexual identity is a 'production' within
   these relations, as I have mentioned in an earlier chapter by reference
   to Rubin's position (see chapter 5, note 13). It is unclear whether Delphy
   recognises this. It would appear that she sees ideology as more than
   subjectivity, but views subjectivity as an aspect of ideology. Similarly she
   seems to describe sexuality generally as (merely) ideological. G. Rubin,
   'The traffic in women: Notes on the 'Political Economy' of Sex', in
   *Toward an Anthropology of Women*, ed. R. Reiter, Monthly Review
   Press, N.Y./London, 1975, p. 167; Delphy, 'A materialist feminism is
   possible', p. 92.
13 Delphy, 'A Materialist Feminism is possible', p. 97. Hales raises a similar
   point to that of Delphy in the field of class analysis. M. Hales, *Living
   Thinkwork: Where Do Labour Processes Come From?*, CSE Books,
   London, 1980, p. 86.
14 Thompson, *The Nature of Work*, p. 177.
15 MacKinnon makes use of certain features of Marxist method to look at
   particular feminist questions. For example, she argues that consent has
   no meaning in rape laws: she attempts to expose what she sees as the
   myth/surface appearance of fairness in sexual relations which belies
   underlying relations of oppression and expropriation. C. MacKinnon,
   'Feminism, Marxism, method, and the state: Toward feminist
   Jurisprudence', in *Feminism and methodology*, ed. S. Harding, Indiana
   University Press/Open University Press, Bloomington, Indiana/Milton
   Keynes, 1987, pp. 141–7; J. Miller, 'Comments on MacKinnon's "Feminism, Marxism, method, and the state," ' *Signs*, vol. 10, no. 1, Autumn
   1984, pp. 171–2.
16 Ferguson states that 'each mode of sex/affective production . . . [has]
   its own distinctive logic of exchange of the human services of sexuality,
   nurturance, and affection, and will therefore differently constitute the
   human nature of its *special product: human children*' [*emphasis added*].
   Ferguson, 'On conceiving motherhood and sexuality', p. 155.
17 ibid., p. 155–6; A. Ferguson, *Blood at the Root: Motherhood, Sexuality
   and Male Dominance*, Pandora Press, London, 1989, p. 83.
18 Delphy, 'A materialist feminism is possible', p. 96–7.
19 This is a common problem, reflecting in part Delphy's view that there
   is at present no elaborated theory of the subject other than rather
   dehistoricised psychological/psychoanalytic models. The problem occurs,
   for example, in Hartsock's work according to Allen. Ferguson has also
   referred to such difficulties in relation to theories of the subject. J. Allen,

Review (Hartsock), *Signs*, vol. 10, no. 3, Spring 1985, pp. 578–9; Ferguson, 'On conceiving motherhood and sexuality', pp. 176, 178–9.

20 See my previous comments in chapter 6 on Chodorow. Mitchell's lack of clarity about the relationship between labour and the psyche is evident in an interview published in 1983. While Mitchell and Rose argue that the psychic never directly reflects the social, they do acknowledge that there is some relation between the symbolic/psychic and referents. However, given that Mitchell and Rose concur with Lacan's view of the eternal character of phallic symbolic power, change within the social cannot alter psychic phallocentrism. This would appear irreconcilable with *any* suggestion of a relation between the social and the psychic, and thus with their own account of the existence of some (undefined) relation. P. Adams and E. Cowie, 'Feminine sexuality: Interview with Juliet Mitchell and Jacqueline Rose, *m/f*, no. 8, 1983, pp. 3–16.

21 I first considered the question of combining the seemingly opposing approaches of Chodorow and Mitchell in C. Beasley, *The Ambiguities of Desire: Patriarchal Subjectivity*, unpublished M.A. thesis, Centre for Contemporary Cultural Studies, University of Birmingham, U.K., 1985, pp. 105–12. These approaches, or combinations of them, do not of course exhaust the possibilities of psychoanalytic frameworks that could be employed in a feminist economics. Ferguson, for example, refers also to the potentialities of other writers in the psychoanalytic tradition such as Luce Irigaray, whose concern with the materiality of feminine sexuality suggests particular insights into the epistemological limits of sexual economyths and the construction of feminist epistemologies dealing with sexual specificity. Ferguson, *Blood at the Root*, pp. 47–51 and 62–4.

22 As Tong has pointed out, though the myth of the Oedipal crime has been criticised, the significance of this myth in social relations is not easily dismissed. This may suggest the continuing usefulness of the notion of the Oedipus complex, even if it requires some reworking to allow for contextual differentiation, as well as the continuing usefulness of employing myth per se to illustrate and explore subjectivity.

23 Chodorow's emphasis on the importance of the female signifier is similar to related theories of womb envy and men's desire for procreativity/immortality/potency. E. Kittay, 'Womb envy: An explanatory concept', and Al-Hibbri, 'Reproduction, mothering, and the origins of patriarchy', pp. 94–128; pp. 81–93; M. O'Brien, *The Politics of Reproduction*, RKP, London/Boston, 1981.

24 This omnipresent singularity is even more open to doubt when it is considered that the notion of the eternal Oedipus complex is drawn from a particular Western family form and sexual order. This remains a problem within psychoanalytic models that focus on the pre-Oedipal period. Tong, *Feminist Thought*, pp. 157–8.

25 Though Irigaray and other 'French feminists' have postulated a psychoanalytic approach which deals with sexual difference, they, no more than Mitchell and Chodorow, do not suggest differentiated myths. E. Grosz, *Sexual Subversions: Three French Feminists*, Allen & Unwin, Sydney, 1989; M. Whitford, *Luce Irigaray: Philosophy in the Feminine*, Routledge, London/N.Y., 1991.

26  It is interesting in this context to consider Segal's semi-autobiographical discussion of the complex relationship between desire/fantasy and the social relations which women consciously seek in their domestic and public lives. Her account suggests no clear 'base' for feminine subjectivity, with its mixture of conscious and unconscious imperatives. L. Segal, 'Sensual uncertainty, or why the clitoris is not enough', in *Sex & Love: New Thoughts on Old Contradictions*, eds. S. Cartledge and J. Ryan, Women's Press, London, 1983, pp. 43 and 45.

27  Ferber's survey, for example, of writings on women and work in the 1980s indicates that the balance of research is heavily on the side of waged labour. M. Ferber, 'Women and work: Issues of the 1980s', *Signs*, vol. 8, no. 2, Winter 1982, passim.

28  D. Ironmonger and E. Sonius, 'Household productive activities', and K. Funder, 'The value of work in marriage', in *Households Work: Productive activities, women and income in the household economy*, ed. D. Ironmonger, Allen & Unwin, Sydney, 1989, p. 18, p. 177; Australian Bureau of Statistics (ABS), 'Measuring unpaid household work: Issues and experimental estimates', ABS Information Paper, Catalogue No. 5236.0, Commonwealth Government, Canberra, 1990, p. 7, Ferber, 'Women and work, p. 280.

29  Hartsock, 'The feminist standpoint', p. 166; Ferguson, 'On conceiving motherhood and sexuality', p. 159. Sayers notes that the distinction between 'work' and 'non-work' is difficult to make even within class analysis, though he appears to underestimate the problem and ignores the fact that what he views as 'leisure' may be defined as work for women. S. Sayers, 'Work, leisure and human needs', *Thesis Eleven*, no. 14., 1986, passim. Ferguson outlines a model developed by Nancy Folbre which enables a distinction between 'work' and 'leisure' in unpaid household activity. However, Folbre's distinction involves referring to 'market equivalents' and 'contextual expectations'. A central theme of this book has been to note the difficulties of methodologies which transpose market categories and distinctions to analysis of domestic labour. Moreover, I remain somewhat uncertain about Folbre's use of 'contextual expectations'. Ferguson provides an example of the meaning of this term in relation to sexual activity: 'if a certain amount of sexual activity is the norm for married couples in a certain historical circumstance, then roughly that amount can be said to make life minimally sexually satisfying for those involved and thus to involve socially necessary labour'. This 'way of dividing work from leisure' implies in my view a conception of shared 'contextual expectations' across the sexes with regard to determining the amount and satisfactions of sexual activity, that is, determining the character of the labour. The methodology is based in the measurement unit of the couple, or at least in the measurement of broadly-based social assumptions about 'the family unit', and tends to subsume women's potentially specific experiences of the distinction between 'work' and 'time off'. See in this context Goodnow's account of (sexually specific) meanings attached to particular labours. Ferguson, *Blood at the Root*, pp. 94–5; J. Goodnow, 'Work in

households: an overview and three studies', in *Households Work*, ed. D. Ironmonger, Allen & Unwin, Sydney, 1989, p. 46.
30 C. Delphy, 'Continuities and discontinuities in marriage and divorce', in *Sexual Divisions and Society: Process and Change*, eds. D. Barker and S. Allen, Tavistock, London, 1976, pp. 77–8.
31 Wives contribute to their husbands' exchange value activities if, for example, their husbands are self-employed, professionals or farmers, ibid.
32 A. Dworkin, *Right-wing Women*, Coward-McCann, N.Y., 1983, pp. 174 and 184. The problem with these models, in my view, is that they depend upon analogies very much derived from the market and labour relations 'outside' the private sphere.
33 A. Rich, 'Compulsory heterosexuality and lesbian existence', *Signs*, vol. 5, no. 4, Summer 1980, pp. 631–90; C. MacKinnon, 'Feminism, Marxism, method, and the state: An agenda for theory', *Signs*, vol. 7, no. 3, Spring 1982, p. 533; Harding, Review (O'Brien), *Signs*, vol. 8, no. 2, Winter 1982, pp. 361–2; Al-Hibri, 'Reproduction, mothering, and the origins of patriarchy', pp. 83–85; Kittay, 'Womb envy: An explanatory concept', pp. 97 and 106–20.
34 Ferguson, 'On conceiving motherhood and sexuality', pp. 155–6 and 158; Ferguson, *Blood at the Root*, p.86, 146–65 and 188–208.
35 See Tong's comments on Jaggar's perspective. Tong, *Feminist Thought*, pp. 128–9.
36 O'Brien says that men try to compensate for their separation from biological continuity: capitalism is a form of 'male continuity principle'. Harding, Review (O'Brien), pp. 361–2. Al-Hibri argues that patriarchal appropriation involves a reversal in which the 'natural provider' becomes the dependent, while the dependent is asserted to be the provider. Men wish to appropriate women's powers of reproduction and capacity to provide. Al-Hibri, 'Reproduction, mothering, and the origins of patriarchy', p. 86. For Kittay, womb envy results in appropriation through intervention in birth and the use of procreative capacities as a metaphor for male activities. Kittay, 'Womb envy: An explanatory concept', pp. 112–13.
37 Ferguson, 'On conceiving motherhood and sexuality', pp. 156–7 and 159–60.
38 Smith's account of distinctions between 'control', 'ownership' and 'possession' is useful here. However, Smith does not appear to recognise as fully as Hartsock that control can have somewhat different meanings in the labour relations of the private sphere. Hartsock notes that women as mothers must deal with the question of excessive control and the gradual relinquishing of control, which is not the same order of problem for waged workers. Smith's analysis is partly limited by the fact that she sees control largely in terms of the power of the dominant group. J. Smith, 'Parenting and property', in *Mothering*, ed. J. Trebilcot, Rowman & Allanheld, Totowa, N.J., 1984, pp. 202–4; Hartsock, 'The feminist standpoint', p. 166.
39 S. Cohen, 'A labour process to nowhere?', *New Left Review*, no. 165, September/October 1987, passim. I do not agree with Cohen's tendency to perceive the post-Braverman labour process debate about control as

a politically and theoretically dangerous diversion from the 'real' significance of ownership and profit in capitalism. Moreover she, like Smith (see note above) views ownership from the perspective of the dominant group. Nevertheless, Cohen correctly points out that 'ownership'—whether this term is applied to class or sex relations—is a critical feature of expropriation and must be analysed alongside notions of control.

40  I have here paraphrased Smith's description of 'ownership', which attempts to take note of the particularities of the private sphere. Smith, 'Parenting and property', p. 203.

41  Hartsock, 'The feminist standpoint', p. 164.

42  K. Spearritt, quoted in G. Reekie, 'Naming Queensland women's history: A bibliographic essay', *Hecate*, vol. 15, no. 2, 1989, p. 97; K. Spearritt, *The Poverty of Protection: Women and Marriage in Colonial Queensland*, unpublished B. A. Honours thesis, History Department, University of Queensland, 1988.

43  Quote from Spearritt in Reekie, 'Naming Queensland women's history: A bibliographic essay', p. 97.

44  Ferguson's work gives some hints about estimating/analysing expropriation in the sense of referring to differential inputs, degrees and kinds of control, outputs, and benefits/satisfactions, but is less clear about how this battery of aspects of expropriation might be measured than Delphy's schematic model. On the other hand, Delphy's account of the dimensions of expropriation is much less detailed and more economistic: clarity is achieved at the expense perhaps of subtlety. ·

45  Funder, 'The value of work in marriage', pp. 174–6 and 182.

46  Eco, *Foucault's Pendulum*, p. 613.

47  A rare exception occurs in Hite's work, but her material focuses on overviews of emotional work described in terms of feelings/attitudes, love, sex and relationships. She deals only marginally with the integration of emotional labour and service tasks like cooking and the complex diversity of differing types of emotional labours. S. Hite, *Women and Love: A Cultural Revolution in Progress*, Penguin, Harmondsworth, 1987.

48  Some of these direct measures, which include analysis/measures of meaning, have been suggested by household work analysts. Goodnow, 'Work in households', pp. 45–6 and 49–50. Hartsock's account of Mainardi's early piece on housework suggests an interesting qualitative measure related to demarcation, transferability and negotiation: 'The measure of your oppression is his resistance.' Hartsock, 'The feminist standpoint', p. 124; P. Mainardi, 'The politics of housework', in *Sisterhood is Powerful*, ed. R. Morgan, Vintage Books, N.Y., 1970, pp. 501–10.

49  S. Dunlevy, 'Wedded bliss better for men than women', *The Advertiser*, November 23, 1990, p.1 (Report based on Australian Bureau of Statistics figures).

50  M. Neave, 'Production and reproduction—does the law recognise the value of women's work?', paper presented at the 27th Australian Legal Convention, Adelaide, 8–12 September, 1991, pp. 10 and 22.

51  Funder, 'The value of work in marriage', pp. 178–9.

52  Neave, 'Production and reproduction', p. 10.
53  'Conjugal compensation', *The News*, October 2, 1990, p. 6. The states of New South Wales, Tasmania and Western Australia have abolished these provisions, which were once available throughout Australia. However,

> the legislation does not specify how loss of ability to work in the home should be compensated . . . There is conflicting case law on whether loss of the ability to perform domestic services should be characterized as loss recoverable by the plaintiff, or a loss suffered by the persons for whom she provided the services, and hence recoverable, if at all, only in an action for loss of consortium.

The states of South Australia and Queensland have extended the right of compensation to women as well as men. This would appear initially to be the best alternative, 'but .this gender–neutral approach [does] not necessarily result in equality of outcome because of the way in which responsibility for domestic work is divided between men and women'. (Neave, 'Production and reproduction', pp. 14, 15–16.) In this context Graycar notes,

> [s]ince the major part of any award for loss of consortium is the services element, the victory, if it is one, is pyrrhic only; most women don't lose their husband's services when the latter are injured: they never had them in the first place. So the net effect [is that] in practical terms, a damages award to a woman for loss of her husband's consortium [is] significantly lower than a man's corresponding award.

(Graycar cited in Neave, 'Production and reproduction', p. 15.)

54  Neave, 'Production and reproduction', p. 14.
55  ibid.
56  Romeyko argued that since the Commonwealth Family Law Act states that the Family Court can make an order to relieve a party from the obligation to perform marital services, this implies that without such an order the obligation remains. Hence he asserted that the state rape-within-marriage legislation of South Australia is invalid because it contravenes the Commonwealth Family Law Act. D. Bevan, 'Rape–in–marriage legal row looms', *The Advertiser*, October 31, 1990, p. 3; R. Ogier, 'One man's fight over rape–in–marriage law', *The Advertiser*, November 1, 1990, p. 2.
57  Seddon asserts that the existence of injunctions to prevent unwanted sexual conduct by a spouse suggests that Family Law assumes consent to provide sexual services in the absence of such injunctions. Consent to provide sexual services in marriage is implied as the norm, against which injunctions come into play as exceptional provisions. (My thanks to Ngaire Naffine for help on this point.) N. Seddon, *Domestic Violence in Australia: The Legal Response*, Federation Press, Annandale, NSW, 1989, p. 40. See also Graycar's discussion of legal assumptions about provision of a range of other services in marriage which arise, for example, in relation to women caring for injured spouses. R. Graycar, 'Love's labour's cost: The High Court decision in Van Gervan v Fenton', *Torts Law Journal*, vol. 1, no. 2, July 1993, pp. 122–36.

58  Funder, 'The value of work in marriage', pp. 173, 176.
59  See Mackie's analysis of Mies' approach. V. Mackie, 'Writing about women in Asia', *Hecate*, vol. 15, no. 2, 1989, pp. 87–8; M. Mies, *Patriarchy and Accumulation on a World Scale: Women in the International Division of Labour*, Zed Press, London/Atlantic Highlands, N.J., 1986. For a more sympathetic reading of Mies's work on Indian women in Narsapur, see C. Mohanty, 'Under Western eyes: Feminist scholarship and colonial discourses', *Feminist Review*, no. 30, Autumn 1988, pp. 73–4.
60  One example here might be the use of cloth nappies as against disposable ones.

## Conclusion

1   M. Ferber and J. Nelson, 'Introduction: The social construction of economics and the social construction of gender', in *Beyond Economic Man: Feminist Theory and Economics*, eds. M. Ferber and J. Nelson, University of Chicago Press, Chicago, 1993, pp. 24–5.
2   M. Barrett, *Women's Oppression Today: Problems in Marxist Feminist Analysis*, Verso/NLB, London, 1980; M. Barrett and M. McIntosh, 'Christine Delphy: Towards a materialist feminism?', *Feminist Review*, no. 1, 1979, pp. 95–106; C. Delphy, 'A materialist feminism is possible', *Feminist Review*, no. 4, 1980, pp. 79–105.
3   L. Bryson, *Welfare and the State*, Macmillan, London, 1992, pp. 40–1
4   S. Harding, 'Conclusion: Epistemological questions', in *Feminism and Methodology: Social Science Issues*, ed. S. Harding, Indiana University Press/Open University Press, Bloomington, Indiana/Milton Keynes, 1987, p. 189
5   S. Harding, 'Why has the sex/gender system become visible only now?', in *Discovering Reality: Feminist Perspectives on Epistemology, Metaphysics, Methodology, and the Philosophy of Science*, eds. S. Harding and M. Hintikka, D. Reidel Pub., Dordrecht, 1983, p. 311.
6   J. Allen, 'Does feminism need a theory of "the state"', in *Playing the State: Australian Feminist Interventions*, ed. S. Watson, Allen & Unwin, Sydney, 1990, p. 34.
7   See contributions in S. Rees et. al eds., *Beyond the Market: Alternatives to Economic Rationalism*, Pluto Press, Leichhardt, NSW, 1993 and D. Horne ed., *The Trouble with Economic Rationalism*, Scribe Pub., Newham, Victoria, 1992.
8   E. Cox, 'The economics of mutual support: A feminist approach', in *Beyond the Market*, eds. S. Rees et al., Pluto Press, Leichhardt, NSW, 1993, pp. 273–4.
9   R. Blank, 'What should mainstream economists learn from feminist theory?', in *Beyond Economic Man*, eds. M. Ferber and J. Nelson, University of Chicago Press, Chicago, 1993, p. 136.
10  ibid., pp. 136–7.
11  ibid., p. 137, 139.
12  ibid., pp. 140.
13  ibid., p. 140, 141

14  ibid., p 133..
15  It is unclear whether Blank's comments regarding the overly general theoretical nature of feminist analyses of economics are meant to apply only to the text she has been asked to discuss or whether she intends them to refer to the whole field of feminist writings on economics and women's labour. She certainly explicitly points out that she has no great knowledge of feminist works. ibid., pp. 133 and 137.
16  V. Beechey, *Unequal Work*, Verso, London, 1987, p. 13.
17  Blank, 'What should mainstream economists learn from feminist theory?', p. 137.
18  Campioni and Gross describe phallocentric epistemologies in terms of their inclination to constitute women and their specificity as dependent categories of analysis, indeed as deficient, variant versions of a masculine norm. See Chapter 3, note 40.
19  Blank, 'What should mainstream economists learn from feminist theory?', p. 137.
20  M. Neave, 'Production and reproduction—Does the Law recognise the value of women's work?', paper presented at the 27th Australian Legal Convention, Adelaide, 8–12 September, 1991; R. Graycar and J. Morgan, *The Hidden Gender of Law*, Federation Press, Leichhardt, NSW, 1990; C. Gilligan, *In a Different Voice: Psychological Theory and Women's Development*, Harvard University Press, Cambridge, Ma., 1982.
21  C. Pateman, *The Sexual Contract*, Polity Press, Cambridge, 1988, p. ix and passim.
22  A. Yeatman, 'Despotism and civil society: The limits of patriarchal citizenship', in *Women's Views of the Political World of Men*, ed. J. Stiehm, Transnational Pub., N.Y., 1984, p. 159.
23  ibid., p. 155.
24  Pateman, *The Sexual Contract*, passim.
25  As I have noted in previous chapters, there is considerable evidence that connections between the household and market economies are limited. This is not to ignore any notion of connection, but simply to offer the thought that Pateman's conception of the 'sexual contract' as underlying and integral to the existence of the public sphere in her analysis of classical political theory may not be directly transferable to the analysis of possible interactions between the public/market and private/household economies. Direct translation of Pateman's approach in the field of economics might rejuvenate, in an inverted fashion, existing economic perspectives depicting an overly functional fit between the operations of the two economies that ignore the ways in which, for example, change in one does not appear to result in anything like equivalent change in the other.
26  E. Keller and J. Flax discussed in S. Harding and M. Hintikka, 'Introduction', in *Discovering Reality*, eds. S. Harding and M. Hintikka, D. Reidel Pub., Dordrecht, 1983, p. 311, p. xvi–xviii.
27  J. Flax, 'Political philosophy and the patriarchal unconscious: A psychoanalytic perspective on epistemology and metaphysics', in *Discovering*

*Reality*, eds. S. Harding and M. Hintikka, D. Reidel Pub., Dordrecht, 1983, p. 245.

28  Harding and Hintikka, 'Introduction', p. xviii.

29  N. Folbre, 'Socialism, feminist and scientific', in *Beyond Economic Man*, eds. M. Ferber and J. Nelson, University of Chicago Press, Chicago, 1993, p. 95.

30  D. Dahlerup, 'Confusing concepts—confusing reality: a theoretical discussion of the patriarchal state', in *Women and the State: The Shifting Boundaries of Public and Private*, ed. A. Sassoon, Hutchinson, London, 1987, p. 106.

31  H. Carby, 'White woman listen! Black feminism and the boundaries of sisterhood', in *The Empire Strikes Back*, eds. Centre for Contemporary Cultural Studies, CCCS/Hutchinson, London, 1982, pp. 212–35; L. Segal, *Is the Future Female?: Troubled Thoughts on Contemporary Feminism*, Virago, London, 1987, p. 63.

# Bibliography

Acker J. and Barry, K. 'Comments on MacKinnon's "Feminism, marxism, method and the state" ', *Signs*, vol. 10, no. 1, Autumn 1984, pp. 175–9

Adams, P. and Cowie, E. 'Feminine sexuality: Interview with Juliet Mitchell and Jacqueline Rose, *m/f*, no. 8, 1983, pp. 3–16

Afshar, H. ed. *Women, Work and Ideology in the Third World*, Tavistock, London/N. Y., 1985

Al-Hibbri, A. 'Reproduction, mothering, and the origins of patriarchy', in *Mothering: Essays in Feminist Theory*, ed. J. Trebilcot, Rowman and Allanheld, Totowa, N.J., 1984, pp. 81–93

Allen, J. 'Marxism and the man question: Some implications of the patriarchy debate', in *Beyond Marxism?: Interventions after Marx*, eds. J. Allen and P. Patton, Intervention Pub., Sydney, 1983, pp. 91–111

——Review (Hartsock), *Signs*, vol. 10, no. 3, Spring 1985, pp. 577–9

——Does Feminism need a theory of "the state"?', in *Playing the State: Australian Feminist Interventions*, ed. S. Watson, Allen & Unwin, Sydney, 1990, pp. 21–37

Allen, J. and Grosz, E. 'Editorial', *Australian Feminist Studies* ('Feminism and the Body'), no. 5, Summer 1987, pp. vii–xi

Australian Bureau of Statistics, 'Measuring unpaid household work: Issues and experimental estimates', ABS Information Paper, Catalogue No. 5236.0, Commonwealth Government, Canberra, 1990

Barrett, M. *Women's Oppression Today: Problems in Marxist Feminist Analysis*, first edition, Verso/NLB, London, 1980

——*Women's Oppression Today: The Marxist/Feminist Encounter*, second edition, Verso/NLB, London, 1988

Barrett, M. and McIntosh, M. 'Christine Delphy: Towards a Materialist Feminism?', *Feminist Review*, no. 1, 1979, pp. 95–106

——*The Anti-social Family*, Verso/NLB, London, 1982

159

Bassi, L. 'Confessions of a feminist economist: Why I haven't yet taught an economics course on women's issues', *Women's Studies Quarterly*, vol. XVIII, nos. 3 & 4, 1990, pp. 42–5

Baxter, J. et al. *Double Take: The Links between Paid and Unpaid Work*, AGPS, Canberra, 1990

Beasley, C. *Educating Rita's Grandmother: The Social Relation of the Sexes and South Australian Curriculum Reform, 1875–1915*, unpublished MEd thesis, Flinders University of South Australia, 1984

——*The Ambiguities of Desire: Patriarchal Subjectivity*, unpublished MA thesis, Centre for Contemporary Cultural Studies, University of Birmingham, U.K., 1985

——'The patriarchy debate: Should we make use of the term "patriarchy" in historical analysis?', *History of Education Review*, vol. 16, no. 2, 1987, pp. 13–20

——*The Sexual Metaphysics of Economics: A Feminist Critique of Postmodernism, Post-Marxism, Marxism and 'Materialist' Feminisms*, PhD thesis, Women's Studies, Flinders University of South Australia, 1991

——'Can the contents of a "tool-box" do the housework?: Considering the uses of a Foucauldian framework for investigating power and women's labour', paper presented at the Australian Sociological Association Conference, Adelaide, December 1992

Beechey, V. *Unequal Work*, Verso, London, 1987

Beilharz, P. 'Marxism and history', *Thesis Eleven*, no. 2, 1981, pp. 7–22

Beneria, L. ed. *Women and Development: The Sexual Division of Labour in Rural Societies*, Praeger Pub., N.Y., 1985

Benhabib, S. and Cornell, D. 'Introduction: Beyond the politics of gender', in *Feminism as Critique: Essays on the Politics of Gender in Late-Capitalist Societies*, eds. S. Benhabib and D. Cornell Polity Press, Cambridge, 1987, pp. 1–15

Bergmann, B. *The Economic Emergence of Women*, Basic Books, N.Y., 1986

——'Feminism and economics', *Women's Studies Quarterly*, vol. XVIII, nos. 3 & 4, 1990, pp. 68–74

——comp. 'Reading lists on women's studies in economics', *Women's Studies Quarterly*, vol. XVIII, nos. 3 &4, 1990, pp. 75–86

Bevan, D. 'Rape-in-marriage legal row looms', *The Advertiser*, October 31, 1990, p. 3

Bittman, M. *Juggling Time: How Australian Families Use Time*, Office of the Status of Women, Department of the Prime Minister and Cabinet, Canberra, 1991

Bland, L. et al., 'Women "inside" and "outside" the relations of production', in *Women Take Issue: Aspects of Women's Subordination*, eds. Centre for Contemporary Cultural Studies (Women's Studies Group), Hutchinson/CCCS, London, 1978, pp. 35–78

Blank, R. 'What should mainstream economists learn from feminist theory?', in *Beyond Economic Man*, eds. M. Ferber and J. Nelson, University of Chicago Press, Chicago, 1993, pp. 133–43

Blau, F. and Ferber, M. *The Economics of Women, Men and Work*, Prentice-Hall, Englewood Cliffs, N.J., 1986

Boserup, E. *Women's Role in Economic Development*, Allen & Unwin, London, 1970

Bradbury, R. 'The Last Circus', in R. Bradbury, *The Toynbee Convector*, Grafton Books, London, 1988

Bryson, L. *Welfare and the State*, Macmillan, London, 1992

Burns, S. *The Household Economy: Its Shape, Origins and Future*, Beacon Press, Boston, 1977

Butler, J. 'Gender, the family and history', *Praxis International*, 7:1, April 1987, pp. 125–30

——*Gender Trouble: Feminism and the Subversion of Identity*, Routledge, N.Y./London, 1990

——'Contingent foundations: Feminism and the question of "postmodernism" ', in *Feminists Theorize the Political*, eds. J. Butler and J. Scott, Routledge, N.Y./London, 1992, pp. 3–21

Campioni, M. and Gross, E. 'Love's labours lost: Marxism and feminism', in Beyond Marxism?: *Interventions after Marx*, eds. J. Allen and P. Patton, Intervention Pub., Sydney, 1983, pp. 113–41

Carby, H. 'White woman listen! Black feminism and the boundaries of sisterhood', in *The Empire Strikes Back*, eds. Centre for Contemporary Cultural Studies, CCCS/Hutchinson, London, 1982, pp. 212–35

Cass, B. 'Rewards for women's work', in *Women, Social Science and Public Policy*, eds. J. Goodnow and C.Pateman, Allen & Unwin, Sydney, 1985, pp. 67–94

Chadeau, A. *Measuring Household Activities: Some International Comparisons*, International Association for Research in Income and Wealth, presented at the 18th General Conference of the Association, Luxembourg, August 1983

Chodorow, N. *The Reproduction of Mothering: Psychoanalysis and the Sociology of Gender*, University of California Press, Berkeley/Los Angeles, 1978

——'Mothering, male dominance and capitalism', in *Capitalist Patriarchy And The Case For Socialist Feminism*, ed. Z. Eisenstein, Monthly Review Press, N.Y./London, 1979, pp. 83–106

Cohen, G. *Karl Marx's Theory of History: A Defence*, Princeton University Press, Princeton, 1978

Cohen, J. M. and Cohen, M. J. *The Penguin Dictionary of Quotations*, Penguin, Harmondsworth, 1960

——*The Penguin Dictionary of Modern Quotations*, second edition, Penguin, Harmondsworth, 1980

Cohen, S. 'A labour process to nowhere?', *New Left Review*, no. 165, September/October 1987, pp. 34–50

'Conjugal compensation', *The News*, October 2, 1990, p. 6.

Connell, R. W. *Gender and Power: Society, the Person and Sexual Politics*, Allen & Unwin, Sydney, 1987

Cox, E. 'The economics of mutual support: A feminist approach, in *Beyond the Market: Alternatives to Economic Rationalism*, ed. S. Rees et al., Pluto Press, Leichhardt, NSW, 1993, pp. 270– 5

Curthoys, A. 'Reply to Rosemary Pringle', *Australian Feminist Studies*, nos. 7/8, Summer 1988, pp. 171–7

Dahlerup, D. 'Confusing concepts—confusing reality: a theoretical discussion of the patriarchal state', in *Women and the State: The Shifting Boundaries of Public and Private*, ed. A. Sassoon, Hutchinson, London, 1987, pp. 93–127

Delphy, C. 'Continuities and Discontinuities in Marriage and Divorce', in *Sexual Divisions and Society: Process and Change*, eds. D. Barker and S. Allen, Tavistock, London, 1976, pp. 76–89

——'A materialist feminism is possible', *Feminist Review*, no. 4, 1980, pp. 79–105

——*Close to Home: A Materialist Analysis of Women's Oppression*, trans. and ed. D. Leonard, Hutchinson, London, 1984

Diamond, I. and Quinby, L. eds. *Feminism and Foucault: Reflections on Resistance*, Northeastern University Press, Boston, 1988

Draper, M. 'Women in the home', in *Households Work: Productive Activities, Women and Income in the Household economy*, ed. D. Ironmonger, Allen & Unwin, Sydney, 1989, pp. 85–92

Dunlevy, S. 'Wedded bliss better for men than women', *The Advertiser*, November 23, 1990, p. 1

Dworkin, A. *Right-wing Women*, Coward–McCann, N.Y., 1983

Eagleton, T. 'Marxism, structuralism, and poststructuralism', *Diacritics*, vol. 15, no. 4, Winter 1985, pp. 2–12

——ed. *Raymond Williams: Critical Perspectives*, Polity Press, Cambridge, 1989

——'Base and Superstructure in Raymond Williams', in *Raymond Williams: Critical Perspectives*, ed. T. Eagleton, Polity Press, Cambridge, 1989, pp. 165–75

Eco, U. *Foucault's Pendulum*, Picador, London, 1989

Editorial, *Feminist Review*, no. 23, June 1986, pp. 3–10

Edwards, A. *Regulation and Repression: The Study of Social Control*, Allen & Unwin, Sydney, 1988

——'The sex/gender distinction: has it outlived its usefulness?', *Australian Feminist Studies*, no. 10, Summer 1989, pp. 1–12

Edwards, M. 'Individual equity and social policy', in *Women, Social Science and Public Policy*, eds. J. Goodnow and C. Pateman, Allen & Unwin, Sydney, 1985, pp. 95–103

——'Commentary', *Households Work: Productive Activities, Women and Income in the Household Economy*, ed. D. Ironmonger, Allen & Unwin, Sydney, 1989, pp. 33–7

Eichler, M. *The Double Standard: A Feminist Critique of Feminist Social Science*, Croom Helm, London, 1980

Eisenstein, Z. 'Developing a theory of capitalist patriarchy and socialist feminism', in *Capitalist Patriarchy and the Case for socialist feminism*, ed. Z. Eisenstein, Monthly Review Press, N.Y./London, 1979, pp. 5–40

Eliade, M. *The Encyclopedia of Religion*, Macmillan, N.Y., 1987

Elshtain, J. *Public Man, Private Woman*, Princeton University Press, Princeton, N.J., 1981

——*Meditations on Modern Political Thought: Masculine/Feminine Themes from Luther to Arendt*, Praeger, N.Y., 1986

Engels, F. *The Origin of the Family, Private Property and the State*, International Pub., N.Y., 1972

England, P. and Farkas, G. *Households, Employment, and Gender: A Social, Economic and Demographic View*, Aldine, N.Y., 1986

Enloe, C. 'Silicon tricks and the two dollar woman', *New Internationalist*, January 1992, pp. 12–14

Ferber, M. 'Women and work: Issues of the 1980s', *Signs*, vol. 8, no. 2, Winter 1982, pp. 273–95

Ferber, M. and Nelson, J. eds., *Beyond Economic Man: Feminist Theory and Economics*, University of Chicago Press, Chicago, 1993

——'Preface', in *Beyond Economic Man: Feminist Theory and Economics*, eds. M.Ferber and J. Nelson, University of Chicago Press, Chicago, 1993

——'Introduction: The social construction of economics and the social construction of gender', in *Beyond Economic Man: Feminist Theory and Economics*, eds. M. Ferber and J. Nelson, University of Chicago Press, Chicago, 1993, pp. 1–22

Ferber, M. and Teiman, M. 'The oldest, the most established, the most quantitative of the social sciences—and the most dominated by men: The impact of feminism on economics', in *Men's Studies Modified: The Impact of Feminism on the Academic Disciplines*, ed. D. Spender, Pegamon Press, N.Y., 1981, pp. 125–39

Ferguson, A. 'On conceiving motherhood and sexuality: A feminist materialist approach', in, *Mothering: Essays in Feminist Theory*, ed. J. Trebilcot, Rowman and Allanheld, Totowa, N.J., 1984, pp. 153–82

——*Blood at the Root: Motherhood, Sexuality and Male Dominance*, Pandora Press, London, 1989

Ferguson, A. et al., 'On "Compulsory heterosexuality and lesbian existence": defining the issues', in *Feminist Theory: A Critique of Ideology*, ed. N. Keohane et al., Harvester Press, Brighton, Sussex, 1982, pp. 147–88

Fernbach, D. 'Introduction', in K. Marx, *The First International and After*, ed. D. Fernbach , vol. 3, Penguin/NLB, Harmondsworth, 1974, pp. 9–72

Flax, J. 'Political philosophy and the patriarchal unconscious: A psychoanalytic perspective on epistemology and metaphysics', in *Discovering Reality: Feminist Perspectives on Epistemology, Metaphysics, Methodology, and the Philosophy of Science*, eds. S. Harding and M. Hintikka, D. Reidel Pub., Dordrecht, 1983, pp. 245–81

——'Postmodernism and gender relations in feminist theory', *Signs*, vol. 12, no. 4, Summer 1987, pp. 621–43

Folbre, N. 'Socialism, feminist and scientific', in *Beyond Economic Man*, eds. M. Ferber and J. Nelson, University of Chicago Press, Chicago, 1993, pp. 94–110

Fonow, M. Review (Jaggar, Vogel, Nicholson), *Signs*, vol. 13, no. 2, Winter 1988, pp. 347–351

Foucault, M. *Language, Counter-Memory, Practice*, ed. D. Bouchard, Cornell University Press, N.Y., 1977

Frankel, B. *Marxian theories of the State: A Critique of Orthodoxy*, Monograph Series 3, Arena Pub., Melbourne, 1978

Fraser, N. 'Towards a discourse ethic of solidarity', *Praxis International*, vol. 5, no. 4, January 1986, pp. 425–29

Fritzell, C. 'On the concept of relative autonomy in educational theory', British *Journal of Sociology of Education*, vol. 8, no. 1, 1987, pp. 23–35

Funder, K. 'The value of work in marriage', in *Households Work: Productive Activities, Women and Income in the Household Economy*, ed. D. Ironmonger, Allen & Unwin, Sydney, 1989, pp. 173–87

Gatens, M. 'A critique of the sex/gender distinction', in *Beyond Marxism?: Interventions after Marx*, eds. J. Allen and P. Patton, Intervention Pub., Leichhardt, NSW, 1983

Giddens, A. *The Constitution of Society*, Polity Press, Cambridge, 1984

Gilligan, C. *In a Different Voice: Psychological Theory and Women's Development*, Harvard University Press, Cambridge, Mass., 1982

Goodnow, J. 'Work in households: an overview and three studies', in *Households Work: Productive Activities, Women and Income in the Household Economy*, ed. D. Ironmonger, Allen & Unwin, Sydney, 1989, pp. 38–58

Gouldner, A. *The Coming Crisis of Western Sociology*, Basic books, N.Y., 1970

Grant, R. and Newland, K. eds. *Gender and International Relations*, Millennium Pub., London, 1991

Grant, W. and Nath, S. *The Politics of Economic Policymaking*, Basil Blackwell, Oxford, 1984

Grass, G. *The Flounder: A Celebration of Life, Food and Sex*, Penguin, Harmondsworth, 1977

Graycar, R. 'Love's labour's cost: The High Court decision in Van Gervan v Fenton', *Torts Law Journal*, vol. 1, no. 2, July 1993, pp. 122–36

Graycar, R. and Morgan, J. *The Hidden Gender of Law*, Federation Press, Leichhardt, NSW, 1990

Greenhut, M. L. *Theory of the Firm in Economic Space*, Meredith Corporation, N.Y., 1970

Gross, E. 'Philosophy, subjectivity and the body: Kristeva and Irigaray, in *Feminist Challenges: Social and Political Theory*, eds. C. Pateman and E. Gross, Allen & Unwin, Sydney, 1986, pp. 125–43

——'Conclusion: What is feminist theory?', in *Feminist Challenges: Social and Political Theory*, eds. C. Pateman and E. Gross, Allen & Unwin, Sydney, 1986, pp. 190–204

Gross, M. and Averill, M. 'Evolution and patriarchal myths of scarcity and competition', in *Discovering Reality: Feminist Perspectives on Epistemology, Metaphysics, Methodology, and Philosophy of Science*, eds. S.Harding and M. Hintikka, D. Reidel Pub., Dordrecht, 1983, pp. 71–95

Grosz, E. 'Notes towards a corporeal feminism', *Australian Feminist Studies*, no. 5, Summer 1987, pp. 1–16

——*Sexual Subversions: Three French Feminists*, Allen & Unwin, Sydney, 1989

Gunew, S. ed. *Feminist Knowledge: Critique and Construct*, Routledge, London/N. Y., 1990

Hales, M. *Living Thinkwork: Where Do Labour Processes Come From?*, CSE Books, London, 1980

Hall, S. 'Rethinking the "base-and-superstructure" metaphor', in *Class,*

*Hegemony and Party*, eds. J. Bloomfield et al., Communist University of London, Lawrence & Wishart, London, 1976, pp. 43–73

——'The "political" and the "economic" in Marx's theory of classes', in *Class and Class Structure*, ed. A. Hunt, Lawrence and Wishart, London, 1977, pp. 15–60

Harding, S. 'What is the real material base of patriarchy and capital?', in *Women and Revolution: A Discussion of the Unhappy Marriage of Marxism and Feminism*, ed. L. Sargent, South End Press, Boston, 1981, pp. 135–63

——Review (O'Brien), *Signs*, vol. 8, no. 2, Winter 1982, pp. 361–63

——'Why has the sex/gender system become visible only now?', in *Discovering Reality: Feminist Perspectives on Epistemology, Metaphysics, Methodology, and Philosophy of Science*, eds. S. Harding and M. Hintikka, D. Reidel Pub., Dordrecht, 1983, pp. 311–24

——'The Instability of the Analytical Categories of Feminist Theory', *Signs*, vol. 11, no. 4, Summer 1986, pp. 645–64

——'Conclusion: Epistemological questions', in *Feminism and Methodology: Social Science Issues*, ed. S. Harding, Indiana University Press/Open University Press, Bloomington, Indiana/Milton Keynes, 1987, pp. 181–90

Harding, S. and Hintikka, M. eds. *Discovering Reality: Feminist Perspectives on Epistemology, Metaphysics, Methodology, and Philosophy of Science*, D. Reidel Pub., Dordrecht, 1983

——'Introduction', in *Discovering Reality: Feminist Perspectives on Epistemology, Metaphysics, Methodology, and the Philosophy of Science*, eds. S. Harding and M. Hintikka, D. Reidel Pub., Dordrecht, 1983, pp. ix–xix

Hartmann, H. 'The historical roots of occupational segregation: Capitalism, patriarchy and job segregation by sex', (Part II), *Signs*, vol. 1, no. 3, Spring 1976, pp. 137–69

——'The unhappy marriage of marxism and feminism: Towards a more progressive union', in *Women and Revolution: A Discussion of the Unhappy Marriage of Marxism and Feminism*, ed. L. Sargent, South End Press, Boston, 1981, pp. 1–41

——'The family as the locus of gender, class and political struggle: The example of housework', in *Feminism and Methodology: Social Science Issues*, ed. S. Harding, Indiana University Press/Open University Press, Bloomington, Indiana/Milton Keynes, 1987, pp. 109–34

Hartsock, N. *Money, Sex, and Power: Toward a Feminist Historical Materialism*, Longman, N.Y., 1983

——'The feminist standpoint: Developing the ground for a specifically feminist historical materialism', in *Feminism and Methodology: Social Science Issues*, ed. S. Harding, Indiana University Press/Open University Press, Bloomington, Indiana/Milton Keynes, 1987, pp. 157–80

——'Foucault on power: A theory for women?', in *Feminism/Postmodernism*, ed, L. Nicholson, Routledge, N.Y., 1990, pp. 157–75

Heckman, S. *Gender and Knowledge: Elements of a Postmodern Feminism*, Polity Press, Cambridge, 1990

Hintikka, M. and Hintikka, J. 'How Can Language Be Sexist?', in *Discov-*

*ering Reality: Feminist Perspectives on Epistemology, Metaphysics, Methodology, and Philosophy of Science*, eds. S. Harding and M. Hintikka, D.Reidel Pub., Dordrecht, 1983, pp. 139–48

Hirsch, M. 'Mothers and Daughters', *Signs*, vol. 7, no. 1, Autumn 1981, pp. 200–22

Hite, S. *Women and Love: A Cultural Revolution in Progress*, Penguin, Harmondsworth, 1987

hooks, b. *Feminist Theory from Margin to Center*, South End Press, Boston, Mass., 1983

——*Talking Back*, South End Press, Boston, Mass., 1989

——*Yearning: Race, Gender, and Cultural Politics*, South End Press, Boston, Mass., 1990

Horne, D. ed., *The Trouble with Economic Rationalism*, Scribe Pub., Newham, Victoria, 1992

Huxley, M. 'Commentary' in *Households Work: Productive Activities, Women and Income in the Household Economy*, ed. D. Ironmonger, Allen & Unwin, Sydney, 1989, pp. 79–82

Hyman, P. 'The use of economic orthodoxy to justify inequality: A feminist critique', in *Feminist Voices: Women's Studies Texts for Aotearoa/New Zealand*, ed. R. Du Plessis, Oxford University Press, Auckland, 1992, pp. 252–65

Ironmonger, D. ed. *Households Work: Productive Activities, Women and Income in the Household Economy*, Allen & Unwin, Sydney, 1989

——'Preface', in *Households Work: Productive Activities, Women and Income in the Household Economy*, ed. D. Ironmonger, Allen & Unwin, Sydney, 1989, pp. ix–xi

——'Households and the household economy', in *Households Work: Productive Activities, Women and Income in the Household Economy*, ed. D. Ironmonger, Allen & Unwin, Sydney, 1989, pp. 3–13

Ironmonger, D. and Sonius, E. 'Household productive activities', in *Households Work: Productive Activities, Women and Income in the Household Economy*, ed. D. Ironmonger, Allen & Unwin, Sydney, 1989, pp. 18–32

Jackson, S. 'Towards a historical sociology of housework: a materialist feminist analysis', *Women's Studies International Forum*, vol. 15, no. 2, pp. 153–72

Jacquette, J. 'Power as ideology: A feminist analysis', in *Women's Views of the Political World of Men*, ed. J. Stiehm, Transnational Pub., N.Y., pp. 9–29

Jaggar, A. *Feminist Politics and Human Nature*, Rowman and Allanheld, Totowa, N.J., 1983

Jamrozik, A. 'The household economy and social class', in *Households Work: Productive Activities, Women and Income in the Household Economy*, ed. D. Ironmonger, Allen & Unwin, Sydney, 1989, pp. 64–78

Jennings, A. 'Public or private? Institutional economics and feminism', in *Beyond Economic Man: Feminist Theory and Economics*, eds. M. Ferber and J. Nelson, University of Chicago Press, Chicago, 1993, p. 111–29

Johnson, C. *The transformation of proletarian consciousness in Marx's theory of revolution*, unpublished MA. Econ. thesis, University of Manchester, 1978

——'Some problems and developments in Marxist feminist theory', in *Working it Out: All Her Labours*, ed. Women and Labour Publications Collective, Hale & Iremonger, Marrickville, NSW, 1984, pp. 121–32

——Review (women and politics), *Australian Feminist Studies*, no. 1, Summer 1985, pp. 139–48

Jones, M. A. *The Australian Welfare State*, Allen & Unwin, Sydney, 1980

Kittay, E. 'Womb Envy: An Explanatory Concept', in *Mothering: Essays in Feminist Theory*, ed. J. Trebilcot, Rowman and Allanheld, Totowa, N.J., 1984, pp. 94–128

Laclau, E. and Mouffe, C. 'Socialist Strategy: Where Next?', *Marxism Today*, vol. 25, no. 1, January 1981, pp. 17–22

——'Post-Marxism without apologies', *New Left Review*, no. 166, November/December 1987, pp. 79–106

Lake, M. Review (Connell), *Thesis Eleven*, no. 23, 1989, pp. 160–65

Larkin, J. and J. Brinkworth, 'Child rebate boost: New bid to pay wives', *The Advertiser*, March 25, 1991, p. 1

Leacock, E. 'History, development, and the division of labour by sex: Implications for organization', *Signs*', vol. 7, no. 2, Winter 1981, pp. 474–91

Lewenhak, S. *The Revaluation of Women's work*, second edition, Earthscan Pub., London, 1992

Lichtman, R. 'Marx's theory of ideology', *Socialist Revolution*, vol. 5, no. 23, April 1975, pp. 45–77

Lipsey, R. G. *An Introduction to Positive Economics*, Weidenfeld & Nicolson, London, 1963

Maas, F. 'Commentary', in *Households Work: Productive Activities, Women and Income in the Household Economy*, ed. D. Ironmonger, Allen & Unwin, Sydney, 1989, pp. 14–17

Mackie, V. 'Writing about women in Asia', *Hecate*, vol. 15, no. 2, 1989, pp. 85–91

MacKinnon, C. 'Feminism, Marxism, method and the state: An agenda for theory', *Signs*, vol. 7, no. 3, Spring 1982, pp. 515–44

——'Reply to Miller, Acker and Barry, Johnson, West, and Gardiner', *Signs*, vol. 10, no. 1, Autumn 1984, pp. 184–88

——'Feminism, Marxism, method, and the state: Toward feminist jurisprudence', in *Feminism and Methodology: Social Science Issues*, ed. S. Harding, Indiana University Press/Open University Press, Bloomington, Indiana/Milton Keynes, 1987, pp. 135–56

Mainardi, P. 'The politics of housework', in *Sisterhood is Powerful*, ed. R. Morgan, Vintage, N.Y., 1970, pp. 501–10

Markus, M. ' "Deconstructed" inequality: Women and the public sphere', *Thesis Eleven*, No. 14, 1986, pp. 124–28

Marx, K. *Capital*, vol. 1, Penguin, Harmondsworth, 1976

Marx, K. and Engels, F. *The German Ideology*, Progress Pub., Moscow, 1976

Mascia-Lees, F. et al. 'The postmodernist turn in anthropology: Cautions from a feminist perspective', *Signs*, vol. 15, no. 1, Autumn 1989, pp. 7–33

168 SEXUAL ECONOMYTHS

McCrate, E. 'Comment on Ferber's "Women and work: Issues of the 1980s" ', *Signs*, vol. 9, no. 2, Winter 1983, pp. 326–30

McLennan, G. 'Philosophy and history: some issues in recent marxist theory', in *Making Histories: Studies in History—Writing and Politics*, eds. R. Johnson et al., Hutchinson/CCCS, London/Birmingham, 1982, pp. 133–52

Mies, M. *Patriarchy and Accumulation on a World Scale: Women in the International Division of Labour*, Zed Books, London/Atlantic Heights, N.J., 1986

Miller, J. 'Comments on MacKinnon's "Feminism, Marxism, method and the state", *Signs*, vol. 10, no. 1, Autumn 1984, pp. 168–75

Millman, M. and Kanter, R. 'Introduction to *Another Voice: Feminist Perspectives on Social Life and Social Science*', in *Feminism and Methodology: Social Science Issues*, ed. S. Harding, Indiana University Press/Open University Press, Bloomington, Indiana/Milton Keynes, 1987, pp. 29–36

Minh-ha, T. 'Difference—A special third world issue', *Feminist Review*, no. 25, 1987, pp. 5–20

Mitchell, J. *Psychoanalysis and Feminism*, Penguin, Harmondsworth, 1975

——'Introduction I', in *Feminine Sexuality: Jacques Lacan and the école Freudienne*, eds. J. Mitchell and J. Rose, trans. J. Rose, Macmillan, London, 1982, pp. 1–26

Mohanty. C. 'Under western eyes: Feminist scholarship and colonial discourses', Feminist Review, no. 30, Autumn 1988, pp. 61–88

Morokvasic, M. 'Fortress Europe and migrant women', *Feminist Review*, no. 39, 1991, pp. 69–84

Morris, L. *The Workings of the Household*, Polity Press, Cambridge, 1990

Morris, M. *The Pirate's Fiancée: Feminism, Reading and Postmodernism*, Verso, London/N. Y., 1988

Mumford, K. Women Working: Economics and Reality, Allen & Unwin, Sydney, 1989

Murphy, J. Review (Turner), *Thesis Eleven*, no. 22, 1989, pp. 128–30

National Women's Consultative Council, 'International Labour Organisation Convention 156: Workers with Family Responsibilities', Commonwealth of Australia, AGPS, Canberra, 1990

Neave, M. 'Production and reproduction—Does the law recognise the value of women's work?', paper presented at the 27th Australian Legal Convention, Adelaide, 8–12 September, 1991

Nicholson, L. ed. *Feminism/Postmodernism*, Routledge, N.Y., 1990

——'Feminism and Marx: Integrating kinship with the Economic', in *Feminism as Critique: Essays on the Politics of Gender in Late Capitalist Societies*, eds. S. Benhabib and D. Cornell, Polity Press, Cambridge, 1987, pp. 16–30

——*Gender and History: The Limits of Social Theory in the Age of the Family*, Columbia University Press, N.Y., 1986

O'Brien, M. *The Politics of Reproduction*, RKP, London/Boston, 1981

O'Connor, J. *The Fiscal Crisis of the State*, St. Martins Press, N.Y., 1973

O'Donnell, C. and Hall, P. *Getting Equal*, Allen & Unwin, Sydney, 1988

Oakley, A. *The Sociology of Housework*, Pantheon, N.Y., 1974

Offe, C. *Disorganized Capitalism: Contemporary Transformations of Work and Politics*, ed. J. Kean, Polity Press, Cambridge, 1985

Office of the Status of Women, *Selected Findings from Juggling Time*, Department of Prime Minister and Cabinet, Canberra, 1991

——'Working Families Issues Kit', Department of Prime Minister and Cabinet, Commonwealth Government, AGPS, Fyshwick, ACTU, 1992

Ogier, R. 'One man's fight over rape-in-marriage law', *The Advertiser*, November 1, 1990, p. 2

Pateman, C. *The Sexual Contract*, Polity Press, Cambridge, 1988

Pateman, C. and Gross, E. eds. *Feminist Challenges: Social and Political Theory*, Allen & Unwin, Sydney, 1986

Pateman, C. and Shanley, M. 'Introduction', in *Feminist Interpretations and Political Theory*, eds. M. Shanley and Pateman, C. Polity Press, Cambridge, 1991, pp. 1–10

Peterson, V. S. ed. *Gendered States: Feminist (Re)Visions of International Relations Theory*, Lynne Rienner, Boulder, Colorado, 1992

Pettman, J., 'Women, nationalism and the state: Towards an international feminist perspective', paper presented for the Australian Women's Studies Association Conference, University of Sydney, 1992

Phillips, A. *Hidden Hands: Women and Economic Policies*, Pluto Press, London, 1983

Power, M. et al., 'Writing women out of the economy', paper prepared for the ANZAAS Centenary Congress, Sydney, May 17, 1986

Poynton, C. 'The privileging of representation and the marginalizing of the interpersonal: a metaphor (and more) for contemporary gender relations', in *Feminine Masculine and Representation*, eds. T. Threadgold and A. Cranny-Francis, Allen & Unwin, Sydney, 1990, pp. 231–55

Pringle, R. ' "Socialist-Feminism" in the eighties: Reply to Curthoys', *Australian Feminist Studies*, no. 6, Autumn 1988, pp. 25–30

——*Secretaries Talk: Sexuality, Power and Work*, Allen & Unwin, Sydney, 1988

Rancière, J. 'On the theory of ideology (the politics of Althusser)', *Radical Philosophy*, no. 7, Spring 1974, pp. 2–15

Rapp, R. 'Anthropology', *Signs*, vol. 4, no. 3, Spring 1979, pp. 497–513

Reekie, G. 'Naming Queensland women's history: A bibliographic essay', *Hecate*, vol. 15, no. 2, 1989, pp. 92–110

Rees, S. et al. eds., *Beyond the Market: Alternatives to Economic Rationalism*, Pluto Press, Leichhardt, NSW, 1993

——'Introduction', in *Beyond the Market: Alternatives to Economic Rationalism*, eds. S. Rees et al., Pluto Press, Leichhardt, NSW, 1993, pp. 7–12

Rich, A. 'Compulsory heterosexuality and lesbian existence', *Signs*, vol. 5, no. 4, Summer 1980, pp. 631–90

——*Of Woman Born: Motherhood as Experience and Institution*, Virago, London, 1983

Rosaldo, M. 'The use and abuse of anthropology: Reflections on feminism and cross-cultural understanding', *Signs*, vol. 5, no. 3, Spring, 1980, pp. 389–417

Rowland, R. 'Reproductive technologies: The final solution to the woman

Question', in *Test-Tube Women: What Future for Motherhood?*, eds. R. Arditti et al., Pandora Press, London/Boston, 1984, pp. 356–69

Rubin, G. 'The traffic in women: Notes on the "political economy" of sex', in *Toward an Anthropology of Women*, ed. R. Reiter, Monthly Review Press, N.Y./London, 1975, pp. 157–210

Ruddick, S. 'Maternal thinking' in *Mothering: Essays in Feminist Theory*, ed. J. Trebilcot, Rowman and Allanheld, Totowa, N.J., 1984, pp. 213–30

——'Preservative love and military destruction: Some reflections on mothering and peace', in *Mothering: Essays in Feminist Theory*, ed. J. Trebilcot, Rowman and Allanheld, Totowa, N.J., 1984, pp. 231–62

Russell, G. *The Changing Role of Fathers*, University of Queensland Press, Brisbane, 1983

Sacks, M. 'Unchanging times: A comparison of the everyday life of Soviet working men and women between 1923 and 1966', *Journal of Marriage and the Family*, vol. 39, November 1977, pp. 793–805

Salleh, K. 'Contribution to the critique of political epistemology', *Thesis Eleven*, no. 8, January 1984, pp. 23–43

Samuelson, P. A. *Foundations of Economic Analysis*, Harvard University Press, Cambridge, 1948

Sawer, M. ed. *Australia and the New Right*, Allen & Unwin, Sydney, 1982

——'Why has the Women's Movement had more influence on Government in Australia than elsewhere?', in *Australia Compared: People, Policies and Politics*, ed. F. Castles, Allen & Unwin, Sydney, 1991, pp. 258–77

Seddon, N. *Domestic Violence in Australia: The Legal Response*, Federation Press, Annandale, NSW, 1989

Segal, L. 'Sensual uncertainty, or why the clitoris is not enough', in *Sex & Love: New Thoughts on Old Contradictions*, eds. S. Cartledge and J. Ryan, Women's Press, London, 1983, pp. 30–47

——*Is The Future Female?: Troubled Thoughts on Contemporary Feminism*, Virago, London, 1987

——*Slow Motion: Changing Masculinities, Changing·Men*, Virago, London, 1990

Seidler, V. *Recreating Sexual Politics: Men, Feminism and Politics*, Routledge, London, 1991

Sharp, R. and Broomhill, R. *Short-Changed: Women and Economic Policies*, Allen & Unwin, Sydney, 1988

Simms, M. and Stone, D. 'Women's policy', in *Hawke and Australian Public Policy: Consensus and Restructuring*, eds. C. Jennett and R. Stewart, Macmillan, South Melbourne, 1990, pp. 294–97

Smith, D. 'Women, class and family', *Socialist Register*, 1983, pp. 1–44

Smith, D. *The Rise and Fall of Monetarism: The Theory and Politics of an Economic Experiment*, Harmondsworth, Penguin, 1987

Smith, J. 'Parenting and property', in *Mothering: Essays in Feminist Theory*, ed. J. Trebilcot, Rowman and Allanheld, Totowa, N.J., 1984, pp. 199–212

Smith, P. 'Domestic labour and Marx's theory of value', in *Feminism and Materialism: Women and Modes of Production*, eds. A. Kuhn and A. Wolpe, RKP, London, 1978, pp. 198–219

Spearritt, K. *The Poverty of Protection: Women and Marriage in Colonial Queensland*, unpublished BA Honours thesis, History Department, University of Queensland, 1988

Stilwell, F. 'Contemporary political economy: Common and contested terrain', *Economic Record*, vol. 64, no.184, March 1988, pp. 14–25

Stretton, H. *Political Essays*, Georgian House, Melbourne, 1987, *The Oxford Dictonary of Quotations*, Second Edition, Oxford University Press, London, 1953.

Theile, B. 'Vanishing acts in social and political thought: Tricks of the trade', in *Feminist Challenges: Social and Political Theory*, eds. C. Pateman and E. Gross, Allen & Unwin, Sydney, 1986, pp. 30–43

Thompson, D. 'The sex/gender distinction: A reconsideration', *Australian Feminist Studies*, no. 10, Summer 1989, pp. 23–31

Thompson, M. 'Comment on Rich's "Compulsory heterosexuality and lesbian existence" ', *Signs*, vol. 6, no. 4, Summer 1981, pp. 790–94

Thompson, P. *The Nature of Work: An Introduction to Debates on the Labour Process*, Macmillan, London, 1983

Thornton, M. 'Commentary', in *Households Work: Productive Activities, Women and Income in the Household Economy*, ed. D. Ironmonger, Allen & Unwin, Sydney, 1989, pp. 59–63

Tong, R. *Feminist Thought: A Comprehensive Introduction*, Unwin Hyman, London, 1989

Touraine, A. 'Endgame', *Thesis Eleven*, no. 23, 1989, pp. 117–30

—— 'Is Sociology still the study of Society?', *Thesis Eleven*, no. 23, 1989, pp. 5–34

Turner, B. *Medical Power and Social Knowledge*, Sage, London, 1987

United Nations Division for Economic and Social Information, Department of Public Information, 'Worsening situation of women will be main issue confronting Commission on the Status of Women', Note No. 22, International Women's Decade, February 13, 1980

Vogel, L. *Marxism and the Oppression of Women: Toward a Unitary Theory*, Rutgers University Press, New Brunswick, N.J., 1984

Waring, M. *Counting for Nothing: What Men Value and What Women are Worth*, Allen & Unwin/Port Nicholson Press, Sydney, 1988

Wheelwright, E. L. 'Are the rich getting richer and the poor poorer? If so, why?', in *Questions for the Nineties*, ed. Gollan, A., Left Book Club, Sydney, 1990, pp. 199–215

Whitford, M. *Luce Irigaray: Philosophy in the Feminine*, Routledge, London/N. Y., 1991

Wilenski, P. *Public Power and Public Administration*, Hale and Iremonger, Sydney, 1986

Yeatman, A. 'Despotism and civil society: The limits of patriarchal citizenship', in *Women's Views of the Political World of Men*, ed. J. Stiehm, Transnational Pub., N.Y., 1984, pp. 153–176

Young, K. et al. eds. *Of Marriage and the Market: Women's subordination internationally and its lessons*, RKP, London, 1984

Young, I. 'Beyond the unhappy marriage: A critique of the dual systems theory', in *Women And Revolution: A Discussion Of The Unhappy*

*Marriage Of Marxism And Feminism*, ed. L. Sargent, South End Press, Boston, 1981, pp. 43–69

——'Is male gender identity the cause of male domination?', in *Mothering: Essays in Feminist Theory*, ed. J. Trebilcot, Rowman and Allanheld, Totowa, N.J., 1984, pp. 129–146

Zaretsky, E. 'Socialism and Feminism I: Capitalism, the family, and personal life, Part I', *Socialist Revolution* 3, nos. 1/2, January–April 1973, pp. 69–125

# Index

accumulation, 85
Adams, Jane, 116
Al-Hibbri, A., 98, 99
alienation, 15–16, 59, 65
Allen, J., 12, 62, 108, 112
altruism, 109, 117
androcentrism, viii, x
appropriation, 17, 22, 24, 87, 98;
    *see also* expropriation
Australia, household work studies,
    49
Australian Bureau of Statistics, 39,
    41, 44, 45

Barrett, Michele: debt to Marxism,
    11; dual systems approach,
    67–8, 83; on economic
    determinism, 12; economic
    epistemology, 108; historical
    functionalism, 88, 142n;
    materialist feminism, vii, viii, 106
base/superstructure model, social
    relations, 3–9, 35, 60–2, 92–4,
    125–8n
Bebel, Auguste, 116
Beechey, V., 112
Bergmann, Barbara, xi
*Beyond Economic Man*, 109

Bittman, Michael, 30, 39, 41
Blank, Rebecca, 109–13
body, the, 124n, 130n
Bradbury, Ray, 75
Broomhill, R., 72, 77
Burns, Scott, 29–30

Campioni, M., 12, 50–1
Canada, household work studies,
    49
*Capital*, 24
capitalism: domesticity before,
    62–3; epistemology, 15;
    patriarchy, 10, 81, 84, 85–6,
    88, 147n; production, 8; sex
    relations, 66; social relations, 7,
    46; unified system approach, 65
Carlyle, Thomas, 38, 44–5
Carroll, Lewis, 57
cathexis, 32
causality, 9
childbirth, 24–5
childcare, 17–18, 22, 25, 30–3,
    61, 94, 102, 103, 135n
children, socialisation of, 31
Chodorow, Nancy, xii, 60–1, 80,
    81–2, 95
class relations: appropriation, 17;